Technical
Writing for Social
Scientists

Technical Writing for

John S. Harris
English
Brigham Young University

SOCIAL

SCIENTISTS

and Reed H. Blake
Sociology
Brigham Young University

Nelson-Hall 🔲 Chicago

Library of Congress Cataloging in Publication Data
Harris, John Sterling, 1929-
  Technical writing for social scientists.
  Includes index.
  1. Social sciences—Authorship. 2. English
language—Technical English. I. Blake, Reed H.,
joint author. II. Title.
H91.H38   808'.066'3   75-20129
ISBN 0-911012-39-7
ISBN 0-8829-362-1 PBK

Contents

Chapter 1

Reading maketh a full man,
conference a ready man, and
writing an exact man.

—Francis Bacon

Introduction

—Purpose

The purpose of this book is to teach a specialized form of writing—technical writing—to social scientists. To have social scientists become technical writers may strike some as an unusual undertaking, and perhaps, naturally so, since for many years technical writing has been associated, not with the social sciences, but with the engineering disciplines and with the scientific and technical report.

—The Need for Better Social Science Writing

In recent years, however, a good many people have come to realize that the techniques employed in the technical report can also be useful in the disciplines that make up the social sciences. This book reflects that thinking.

There is a field of science known as the physical sciences, whose unit of analysis is the atom; a field of science known as the biological sciences, whose unit of analysis is the cell; and a field of science known as the social sciences, whose unit of analysis is social man. What all have in common is the utilization of the scientific method. The exactness of the scientific method provides a model for the analysis and the approach of this book. This book maintains that writing in the fields of social science should be an experience in the rhetoric of the scientific method, one in which the author does more than just write about scientific subjects, but strives for prose as rigorously objective and clear as the science that it serves.

—Technical Writing and the Social Sciences

Viewed this way, the writing purpose of the social scientist is the same as that of the engineer: clear, concise writing. The end goal of social science writing is the same as that of the engineer: a technical report. For both there is a need not only to convey information, but also to convey information to a carefully defined live reader.

—The Philosophy of Technical Writing

In technical writing, the concern is not only with writing that can be understood but also with writing that cannot be misunderstood. Clarity is more important than beauty. Writing with precision—that is, with as much precision as an imprecise language will allow—is the major consideration. Further, the stress is upon writing that complements the pragmatic and empirical world of the scientist. Emphasized here is the value of statement, amplification, recapitulation: an internal outline for maintaining the rhetorical and scientific balance between the general statement and the specific instance.

Technical writing, we said, is concerned not only with writing that can be understood but also with writing that cannot be *misunderstood*. This need to avoid misunderstanding imposes an even greater challenge for the social scientist than for the physical or biological scientist. Writing not to be misunderstood is always on the mind of the social scientist as a technical writer, but he cannot work under the conditions of one word–one meaning, one meaning–one word, as some physical and biological scientists are able to do. Because of the nature of social science material, the writer must work under the handicap of one word–many meanings, many meanings–one word.[1]

A popular tongue-in-cheek slogan, displayed in offices, reflects on the ambiguities of language. It reads:

> I know that you believe that you understand
> what you think I said, but I am not sure you
> realize that what you heard is not what I meant.

This apothegm touches upon one of the major problems of the technical writer—that his primary tool in communicating is words—and words often betray him. Using the familiar social science term *behavior* and its common synonyms, Ralph Borsodi ably points to the unavoidable hardship faced by the social science writer:

> The 1938 *Webster's New International
> Dictionary* gives to the word *action* seventeen
> meanings; to the word *act,* sixteen meanings; to the
> word *deed,* eight meanings; to the word *conduct,*
> twenty meanings; and to the words *behave* and
> *behavior,* eight. This is a total of sixty-nine
> meanings, including the duplications of course.
> Linguistically we are confronted with the problem of
> making clear which of these we are referring to
> with one of five synonymous words with meanings
> for each ranging between eight and twenty.
> Roget's *Thesaurus of English Words and
> Phrases* includes seventy-five words which it lists for
> the purpose of expressing the idea of action by
> human beings, and then twelve whole classes of
> words under each of which there is a similarly long
> list of words. If now the number of meanings for
> which all these words are used, and the number of
> words which are used to refer to the same meaning
> are taken into account, the difficulty in the precise
> expression and communication of truths by means
> of language becomes apparent. Mathematicians
> labour under no such difficulty. When a

> mathematician uses the semantheme three, it hasn't
> sixty-nine meanings; it has only one. . . . When a
> physicist or a chemist uses the semantheme
> *hydrogen,* it doesn't refer to sixty-nine elements; it
> refers to only one. And when a biologist uses the
> word *homo sapiens,* it doesn't refer to sixty-nine
> species of living organisms; it refers to only one.[2]

Despite the handicaps of his language, the social scientist still must write. Further, the need for the social scientist to write well has never been greater than it is today. The work of the social scientist as a teacher, scholar, researcher, and social mover, together with the growing need of society and its social problems, generates more teachers, more scholars, more researchers, and more social workers. The standard mode of communication in this vast and growing body is the written word. In addition, the social sciences are now recognized, particularly by government, as they have never been before. This recognition calls for additional facility in writing, as the social scientist communicates to readers unfamiliar with his technical terms. Beyond this, the social scientist, because of the vast amount of knowledge he can bring to bear on his social environment, needs to communicate more effectively with laymen in his society. Unfortunately, the social scientist has been moving away from—rather than toward—such a goal. As Pulitzer Prize historian Barbara Tuchman has said:

> Let us be aware of the plight of our colleagues,
> the behavioral scientists, who by use of a
> proliferating jargon have painted themselves into
> a corner—or isolation ward—of unintelligibility.
> They know what they mean but no one else
> does. . . . Their condition might be pitied if one
> did not suspect it was deliberate. Their retreat into
> the arcane is meant to set them apart from the

great unlearned, to mark their possession of some
unshared, unsharable expertise. No matter how
illuminating their discoveries, if the behavioral
scientists write only to be understood by one another,
they must come to the end of the Mandarins.
Communication, after all, is what language was
invented for.[3]

This reluctance (or inability) to communicate with
others outside their own diciplines has brought much scorn
upon social scientists. In 1947, Samuel Williamson—
from his observation of social science writing to that date
—lampooned the social scientists when he published his
six guide rules for social science writers. Unfortunately,
the proposed "rules" may be more applicable today than
they were nearly three decades ago. Williamson wrote:
1. Never use a short word when you can think
   of a long one.
2. Never use one word when you can use two
   or more.
3. Put one-syllable thoughts into polysyllabic
   terms.
4. Put the obvious in terms of the unintelligible.
5. Announce what you are going to say before
   you say it.
6. Defend your style as "scientific."[4]
Of these "rules," which we blush to admit seem still
to be widely followed, only number five is defensible.
The concern over writing in the social sciences has
not been solely that of the reader. Social scientists, too,
within the various disciplines have been alarmed over the
trend toward noncommunication. Sociologist Hans L.
Zetterberg confirms that for his discipline:

> Our personal conviction is very much in
> favor of having a sociological vocabulary that, in the

main, is understood by most every educated person.
... Unfortunately, the opposite tendency prevails
at present. Instead of speaking of "equal rights,"
some sociologists have learned to say "universalistic
standards," or instead of speaking of the
"familiarity" that prevails in some social relations,
sociologists have learned to say that "diffuseness"
characterizes the relations, and so forth, almost
ad nauseum.[5]

—The Approach of This Book

It is the thesis of this book that the fault of social
science writing lies not in its special vocabularies; rather,
the trouble is that too few social scientists take enough
care with words *outside* their special languages. It is in
this area that the principles of technical writing have
special import to the social sciences.

This lack of communication skill is partly the fault
of social science education. Traditionally, social science
curriculums have shunned, where other disciplines have
not, the inclusion of a required technical writing course.
Subsequently, students in the social disciplines, too, in
working out their academic programs, have ignored writing courses beyond those required for graduation. As a
result, except for those few who would distinguish themselves in any discipline, social scientists admit to an
uneasiness when putting thoughts on paper.

Yet, in a final analysis, the social scientist should
be on familiar ground when engaged in the writing
process, for language is a social phenomenon, an activity
engaged in by people. In this realm, the social scientist
should be at ease. What is it, then, about the communication process that makes him uneasy? His unrest usually
comes from his insecurity in organizing his thoughts and
in writing mechanics.

Yet, even here, the social scientist is not without some basic preparation. The popular adage tells us that "practice makes perfect." This has some application to writing, for experience has proven that the more one writes, the more one tends to write well. But writing entails more than simply putting symbols on paper. It entails a special kind of organized thinking which in time is put to paper. There are those who say that the man who wants to be a social science scholar cannot also be a social science writer—contending that the scholar cannot hope to find the time to serve the million-word apprenticeship that good writing entails—to those people we would answer: the person who is a scholar has much of the necessary skill for effective writing in the ability to organize his thinking that made him a scholar. Just as the scholar is recognized for his precision, so too is scholarly writing recognized for its precision—and precision is a result of sound organization. Technical writing is indeed the rhetoric of the scientific method.

—Summary

These have been general statements about what is and what isn't technical writing. The specific statements about technical writing, particularly in the social sciences, are what this book is about. After he reads this book, then, what can the social scientist as technical writer expect?

1. He can expect to organize, interpret, and report data objectively.
2. He can expect to write correct and clear reports for consumption within and outside his discipline.

The ability to do these tasks should prepare the social science writer to do well what he does so much of—write. (While the beginning scientist is aware of his professional obligation to communicate his research findings, few are aware of the great amount of time that will also be spent in writing things other than research findings. He will be writing proposals, instructions, lectures, and book reviews in addition to writing articles, monographs, and books.)

For the social scientist, the strength of this book, we feel, lies in its dedication to the principal requirements that make him a good technical writer. It is not an attempt to survey the whole field; rather, it concentrates upon the fundamentals, and when examples are used they are drawn from the social sciences.

Chapter 2

The secret of success
is constancy of purpose.

—Disraeli

First
Considerations

Before beginning to write any paper, article, or report, you should first determine who is going to read it and why. You should also, for your own sake and your reader's, give the reader a clear overview of what you intend to cover and how you intend to cover it. This chapter will discuss the problems of adapting the paper to the needs of the intended reader, tailoring the report to suit the planned purpose, and giving the reader the needed preliminary overview of the scope and pattern of development.

—The Reader

Every piece of educated writing should be planned for a particular reader or audience. Who that audience is will determine what is said and how it is said.

When you write a letter, you automatically adjust the form and content of your letter to suit your correspondent. Obviously there is a difference in the letter you write to a senator and to a sweetheart, to a fishing companion and to a clergyman, to a grandmother and to a younger, teen-age brother, to a subordinate and to a superior. A letter that fails to take into account the individuality of the intended recipient will fail to get the kind of effect that you want. Such individual differences must also be considered in an academic paper or scientific report or article. Thus, you should know as much as possible about your intended reader before you begin to write. Consider this:

Age: The generation gap does not exist only across the line of thirty.

Education: What degrees does the man hold? What was his specialization? Where did he study?

Position: What does he do now? What experience has he had with your subject? Is he your colleage, subordinate, or superior?

Political and Social Views: Like race and religion, these can influence strongly or subtly a person's attitudes and understandings of subjects—especially in the social sciences.

The more of these factors the writer can know and consider in the writing of the paper, the better he can make his text fit what the reader wants and needs to know. Of course such things cannot always be known. A given paper may have to reach an audience that is quite varied. In such a situation, the writer would be wise to make a hypothetical reader his intended audience. Such a reader should have average characteristics for the group, but those characteristics should still be very specific. The writer would be wise to write down on a card those characteristics of age, education, position, etc., and to keep that card in front of him at all times during the writing.

A careful consideration of the intended audience is even important in such academic writing as doctoral dissertations, as William Riley Parker makes clear in the *MLA Style Sheet*:

> The effectiveness of all writing, including scholarly writing, depends in part on making certain assumptions and not making others about the interest and previous knowledge of the reader. What assumptions you may justifiably make depends, of course, on the *nature and extent* of your intended

audience, and *you should not begin writing* your
thesis until this problem has been carefully
considered and discussed with the supervisor
of your thesis.[1]

*No other factors are more important to the effec-
tiveness of your writing than careful evaluation of your
intended reader and tailoring of your material to fit him.*

—*Purpose*

Of course you do not just write to a specific reader,
you write to him for a reason. In the academic world,
you usually want to inform him that you have found out
certain information or that you have mastered certain
facts or concepts. As a result, you may want him to be
able to use that information as part of his body of knowledge or you may want him to grant you a grade or an
advanced degree or to recognize you as a scholar.

In the world outside of the university, you may want
your reader to grant you research funds, to vote for or
against a proposed law, or to develop a program to assist
drug users, unwed mothers, or non-English-speaking immigrants.

What you know that the intended reader will do
with the information or what you want him to do with
it will affect what you put into the report. Of course, you
cannot always predict the purposes to which a given
report on, say, "Distribution of Non-White Population
in Greater Philadelphia," will be put. But there is usually
a purpose in making the study that goes beyond the need
for the researcher's grade or salary. That purpose should
control what you say and how you say it in the report,
regardless of how the report is used by later researchers
who come across it.

If the purposes for the report are indefinite or multiple, the same considerations apply as for audience analysis. That is, determine an expedient or arbitrary purpose for the report and write as if it really had that purpose.

In the physical sciences many believe that "pure" science or research such as that associated with chemistry or physics is better (and some moral value is often assumed) than applied science or research such as that done in chemical or mechanical engineering. The same opinion also occurs in the social sciences. Without getting into the controversy of the relative values of pure and applied research, we have noted that the *writing* associated with applied research is generally better. The probable reason for this is that the report of applied research usually has a clearer purpose.

Whether your purpose in writing is implicit in your material or is arbitrarily imposed solely to shape the writing, it is wise to write that purpose down—just as you write down the intended reader and his characteristics—and keep that purpose before you during the entire writing process.

—Scope

We remember the student who handed in the proposed research topic "The World, As It Is Today." We chuckled at his naiveté in thinking that he could cover his topic in ten pages. But we have also read reports by academic professionals who had the same kind of problem with the scope of their reports.

Obviously, the intended reader and the stated purpose for the report will largely dictate how broad and how deep the coverage of the report will be. Still, researchers often want to include in the report everything they have found out—even the information that is irrelevant. If the writer includes in his planning of the scope of his coverage the limitations on time period, geographical area, audience, and the discipline (in the academic sense), he will not go too far astray.

But not only does the theoretical scope need to be determined; that scope also needs to be stated clearly in several places:

1. It should be stated clearly in the title.
2. It should be stated on that same card with the intended reader and purpose.
3. It should be part of the introduction.

Stating in the title what the paper or report will cover will certainly help the subsequent reader who is looking for information on the subject. He can immediately know from the title alone whether the report is likely to be on the subject he wants to know about. Further, making the title an exact statement of what the report will cover aids the cataloger or librarian. If there are strong reasons for adopting an "artistic" or prestigious title that does not describe the subject exactly, then that title may be applied at a later time, but it is helpful to have the working title purely descriptive.

The descriptive or working title also needs to be on that card with the intended reader and purpose of the report and for the same reason. It is a controlling factor on what to include and what not to include.

The scope also needs to be stated in the introduction. Here the statement can be more detailed than in the title; it can here clearly let the reader know if he wants or needs to read more. (Remember that few readers are going to read social science reports for amusement. Rather, they will read one report or many—or parts of one or many—solely to find specific information.)

—*Plan of Development*

Earlier we said that one of Williamson's "rules" for writing has validity. That rule was "Announce what you are going to say before you say it." Such apparent repetition may irritate the reader who has been accustomed to novels and personal essays, but *properly used* it can be a real aid to the writer and reader of the report.

A pattern of rhetoric as old as the ancient Greeks is *Statement, Amplification, Recapitulation.* Following such a pattern, the writer announces his subject, then he discusses it in detail, then he summarizes or draws conclusions. This pattern lets the reader know what will be covered and in what order before he begins to read. In this way, his comprehension is enhanced. The pattern also serves as a guide to the writer.

—*Incorporation of Reader, Purpose, Scope, and Plan into the Introduction*

In this chapter we have talked about the intended reader, the purpose of the paper, the scope, and the plan of development. But these are not just topics to be considered before writing a paper; they are also items to be included in the introduction to the paper. Thus the introduction can very well begin:

Reader: The following report is intended for . . .
Purpose: The purpose of this report is to . . .
Scope: The report will cover . . ., but will not cover . . .
Plan: First the report will discuss . . ., then it will discuss . . ., following that is . . . and . . .

Obviously such a beginning is more utilitarian than pretty, but alterations can be made later to the basic structure to make it more graceful. Nor is the order of treatment of Reader, Purpose, Scope, and Plan a sacred one. The order in which they are mentioned may be altered to suit the circumstances. Also the four items may overlap considerably, since the statement of purpose may also imply who the intended reader is and the statement of plan of development might also include the scope.

The report will be clearer if Reader, Purpose, Scope, and Plan, whether they are implicit or stated, are understood by the end of the first few introductory paragraphs, and with these, as with many other parts of a technical report, it is better to err on the side of being too explicit and too clear.

Chapter **3**

A problem well put is half solved.

—John Dewey

Writing the Sentence and Paragraph

This chapter is concerned with the basic unit of writing, the sentence. It is also concerned with the punctuation of sentences and the organization of sentences into a logical unit, the paragraph.

It is in the sentence that accuracy is achieved—in the selection of the right word and the arrangement of these right words (syntax). The problems of sentence structure lead inevitably to a consideration of punctuation, since punctuation functions in a sentence much as do road signs on the highway. Without proper punctuation, the sentence cannot attain the precision demanded in technical writing. Finally, it is in the paragraph that accurate sentences are organized to promote meaning. This is achieved in paragraph continuity which presents ideas in their proper sequence.

—The Sentence

What is a sentence? Students in grammar school memorize such definitions of sentences as: *A sentence is a group of words that make up a complete thought*—a definition that would be more helpful if it said something about how big a group of words or gave some definition of what is meant by a thought, complete or incomplete. As the definition stands, it can also apply to a paragraph, a chapter, or a volume. Another definition frequently learned is: *A sentence is a group of words having a subject and a predicate*—a definition of some use to anyone who knows what subjects and predicates are, but does not know what a sentence is.

A problem with definitions of sentences is that most of them are designed to be used on sentences that are already produced and awaiting analysis. Our purpose is somewhat different. We are looking for a definition useful to the writer who faces the problem of taking a body of material and transmitting it into sentences.

For our purposes a sentence can be defined as: *A kind of verbal equation used to show a relationship that exists in the universe whether or not anyone talks or writes about it.*

Thus, if a small white dog walks behind a bay mare and she does not like small white dogs, a rather violent action is likely to happen. That action might result in a sentence about the relationship of the things involved, which are:

A horse
A dog
An act of kicking
The degree of force of the action

The writer might choose the dog as the subject:

The dog was kicked by the horse forcibly.

Or the act of kicking might be the subject:

Kicking was the action which was performed forcibly by the horse upon the dog.

Or it might be the degree of force:

Considerable force was expended upon the dog by the horse in the kicking of the former by the latter.

Or the horse might be the subject:

The horse kicked the dog forcibly.

Of course, the last sentence is the clearest. But why? The answer lies back in our definition: *A sentence is a kind of verbal equation used to show a relationship that exists in the universe whether or not anyone talks or writes about it.* And we may now add: *The sentence is likely to be clearest when the grammatical relationship matches the physical relationship.* In our barnyard drama the physical actor was the horse. In the clearest sentence she became the grammatical actor or subject of the sentence. Her action in the physical world became the action part of the sentence, the verb. The recipient of the physical action, the dog, became the grammatical recipient or direct object. And the degree of the action, *forcibly,* became a modifier of the verb.

Usually, the tortured, unclear sentence is one that violates the advice that the grammatical relationship should match the physical relationship. In this sentence:

For the settlement, troubles of a serious nature were had from the first due to the Navaho Indians and their unfriendliness,

the writer has chosen the wrong subject for his sentence. In short, he has failed to find the "horse," that is, the thing that is the actor and that should be the grammatical subject. If he started with the "Indians," he would produce something like:

The Navaho Indians were unfriendly from the first, and this caused serious trouble for the settlement.

Clearly this is an improvement, but starting with the "unfriendliness" would produce an even better focused sentence:

> *From the first, the unfriendliness of the Navaho Indians caused serious trouble for the settlement.*

Such wrong choices of subject may result in such confusion as:

> *In terms of the individual with his relationship to the group, one result that may occur if the individual is unable to make an identification is alienation.*

If the "horse" is seen to be "alienation," then:

> *Alienation may result from the individual's inability to feel that he is part of the group.*

The sentences shown so far have been what grammarians call the SVO or Subject-Verb-Object pattern, or the pattern of actor-action-recipient. A similar pattern eliminates the recipient:

> S   V
> *The horse kicked.*

or

>   S
> *Despite their tremendous power, all glaciers,*
>   V
> *alpine or continental, move slowly.*

The third pattern, rather than showing the action of one thing on another, shows instead some kind of equivalency:

> *The boy is a student.*

This is essentially an equation of:

$$X = X'$$

We are showing that, within this situation, *boy* is the equivalent of *student*.

The verb for such a sentence is called a linking or copulative verb, and it nearly always shows some equivalency, though sometimes with the added dimension of time:

$$X =$$

By the end of the '20s the Freudian theory had

$$X'$$

become the dominant doctrine for all psychological interpretation.

Sometimes the verb may seem to equate a thing with a quality, as it does when a predicate adjective is used:

The house is white

But the situation is ultimately just a shorter way of saying:

$$X = \qquad X'$$

This house is (a) white (house).

Such sentences may often be eliminated simply by putting the adjective in front of the noun, when the noun occurs in the previous or following sentence:

The white house stands on top of the hill.

The SVO, SV, and $X = X'$ patterns are used in nearly all English sentences.

What little is known of the ways we compose sentences in our minds is largely conjecture, but what probably happens is that we perceive the situation and then choose one of the basic sentence equations to fit it. Then we plug in the integers. A failure (a tangled or awkward sentence) may result from choosing the wrong equation or from using the wrong integer, or from both. In speech, such failures are not surprising when we may produce ten sentences a minute, but failures should occur less often in writing, since writing can be produced more deliberately. Still, the language is flexible enough to allow a speaker or writer partially to extricate himself even if he starts the sentence with a wrong choice, though the result is often not completely satisfactory. A better answer, in writing at least, is to go back and recast the sentence, consciously checking to see if the equation is the best one for the material and if each of the integers—particularly the subject—is the best choice.

—More Complicated Patterns

Writing, of course, consists of more than simple, short sentences. The demands of variety, coherence, and emphasis require more sophisticated sentence patterns. How to form these more sophisticated sentences is our next concern.

A small child might relate an event in her life like this:

> Aunt Ellen and I went to the zoo. And we
> saw some monkeys and they were funny and I had
> some ice cream and I saw the elephants and I
> was scared and they were big and I fed them some
> of my peanuts. And we went to the zoo on the
> bus and I had an orangeade and we saw the giraffes
> and I swung on the swings and the lions looked
> mean and we came home in a taxi and I got sick.

Here the child has connected the basic sentence patterns together with the most elementary of connectives—*and*. Under the circumstances the connection is hardly better than none at all. At best, *and* can show such simple relationships as coexistence or a sequential time pattern, but the little girl's narrative does not even allow the reader to understand the time sequence, since she has plainly altered the order of events (doubtless without realizing it).

Connectives of various kinds allow the writer to show rather complex and subtle relationships between what are initially separate thoughts. Besides coexistence, these connectives can show that one idea is the result of another; that two ideas are wholly or partially contradictory; that one idea came about before, during, or after another; or that apparent disagreements between two ideas are only an illusion. Examine what connectives can do to the zoo narrative:

> *When* Aunt Ellen and I went to the zoo, we
> went on the bus. *After* arriving, we saw the elephants,
> *but since* they were big, I was scared. *Nevertheless,*
> I fed them some of my peanuts. *Next* we saw the
> monkeys and they were funny. *Then* I had some
> ice cream. *After that* we saw the giraffes. *Before* I
> swung on the swings, I had an orangeade. *As a result,*
> I got sick. *Therefore* we came home in a taxi.

Of course no small child talks like this. The child does not make the connections between sentence elements because the child does not see the connections between the ideas or events from which the sentences are drawn. Recognition of such relationships as cause and effect, sequence, contradiction, special placement, etc., is the first essential to connecting thoughts. After such recognition, addition of the connective words to show the same relationships is nearly a mechanical process. The following list of words for continuity and transition suggests some of the possibilities for connecting sentence elements.

CONTINUITY WORDS

For Continuation or Sequence	*For Time*	*For Indicating Purpose*
Again	At length	
And	Immediately	To this end
And then	thereafter	For this purpose
Besides	Soon	With this object
Equally	After a few hours	So
important	Afterwards	
Finally	Finally	
Further	Then	
Furthermore	Later	
Moreover	Previously	
Nor	Formerly	
Too	First, second, etc.	
Next	Next	
First, second	And then	
etc.		
Last		
What's more		

For Simultaneous Discussion

For Introducing Consequence	*For Summation*	*For Time Alone*
To this end	In brief	Simultaneously
For this purpose	In short	Meanwhile
With this object	As I have said	In the meantime
So	As has been noted	Meantime
	In other words	

For Emphasis	*For Citing Concrete Instances*
Obviously	For instance
In fact	For example
As a matter of fact	To demonstrate
Indeed	To illustrate
In any case	As an illustration
In any event	
That is	

For Contrast and Paradox

But	On the contrary	Nevertheless
Yet	After all	Notwithstanding
And yet	For all that	On the other hand
However	In contrast	Although this is true
Still	At the same time	While this is true
Nonetheless	Although	Conversely

For Introducing Conclusions

Hence	Thereupon	Summing up
Therefore	Whereupon	To conclude
Accordingly	As a result	In conclusion
Consequently	In brief	In summation
Thus	On the whole	

—Sentence Punctuation

Punctuation marks also convey meaning to the reader. The able handling of punctuation makes the sentence not only clearer but also more interesting to read. The following rules are certainly not intended to be definitive, but they do review basic punctuation marks.

1. The *comma* should be used to:

    A. Separate main clauses (elements that could stand alone as sentences) joined by a coordinate conjunction (and, but, or, nor, for, yet).

        Congress duly passed the enabling legislation, but it failed to appropriate money to pay for the new programs.

    B. Separate an introductory adverbial clause (or a long phrase) from the main clause.

        When nations are in fundamental and long-standing economic competition, war is a specious solution.

C. Separate items in a series (and coordinate adjectives modifying the same noun).

More important than poverty itself are ignorance, apathy, and an eternal sense of defeat. (Note here that the comma before *and* is often omitted, especially in journalism. We retain it in such use because it is sometimes essential to clarity. For most circumstances, however, either form is correct, but you should be consistent.)

D. Set off nonrestrictive and other parenthetical elements.

Consequently, the field, or life space, which influences an individual is described, not in "objective," physical terms, but in the way it exists for that person at that time.[1]

2. The *semicolon* should be used to:

A. Separate two independent clauses not connected by a coordinate conjunction.

Dewey led in the polls and was the predicted winner in the press; Truman won. (Note: we suggest that this construction be used sparingly. One of the words from the previous list of continuity words can make the relationship between the two clauses clear without seeming to be consciously arty.)

B. Separate long coordinate clauses that contain commas.

On the same work crew there were three Italian immigrants, who had been in the United States for forty years but still spoke English brokenly; two middle-aged men

*from the South, neither of whom had fin-
ished grade school; and two graduate school
dropouts, both of whom were discouraged
about the Ph.D. job market.*

C. Separate elements of a series when one or
more of the elements contains commas.

*The applicant must have the following qual-
ities: a master's degree from an accredited
school; five years' work experience, at least
two of which should be in administration;
excellent health; and a willingness to work
for a moderate salary.*

3. The *colon* should be used to:

A. Indicate a following enumeration or explanation.

(Note the previous example.) Such a list
or enumeration can often be made much
easier to read if it is set up as a column with
an introduction rather than in a paragraph.

*The applicant must have the following
qualities:*

    *a. A master's degree from an accred-
ited university;*

    *b. Five years' work experience (two of
which should be in administration);*

    *c. A willingness to work for a moder-
ate salary.*

B. Introduce a long quotation.

*Scharhorst's reply to the charge of self-
interest was written in a letter published in
the* Courant *on November 22:*

    I stand accused by my opponent of

purchasing the land with the hope of re-selling it to the county, but the record will show that the original purchase. . . . .

4. The *dash* should be used to:
   A. Indicate a sharp break in thought and construction.

   "Yes, we understand," they nodded—though what they understood the teacher didn't know.

   B. Set off parenthetical expressions that demand unusual emphasis.

   The virtues that Spencer and Sumner preached—personal providence, family loyalty and family responsibility, hard work, careful management, and proud self-sufficiency—were middle-class virtues.[2]

   C. Introduce a summary of a preceding list.

   Integrity, decisiveness, intelligence, charisma —these are the qualities the electorate demands. (Note that this construction, which is rather the reverse of the formal list introduced by a colon, is not common and should be used sparingly.)

5. The *hyphen* should be used to:
   A. Mark the division of a word at the end of a line.

   When a word is too long to be written without running into the right-hand margin, it should be divided. (Always check the dictionary on proper division of each word. The memory and the ear are notoriously unreliable here.)

B. Connect unit modifiers.

The rapid settlement of the trans-Allegheny region during the thirty years following the Revolution revealed a crucial ambiguity in the conception of agriculture held by Jefferson and Crevecoeur.[3]

6. *Quotation marks* should be used to:
   A. Set off the name of a published article.

   The following year he published an article, "Principles of Population Control in a Democracy," which so incensed the regents that he was called before a hearing board.

   B. Set off the words of another writer or speaker when they are quoted exactly.

   Murphy's second law, "If a number of things can go wrong, the thing that will cause the greatest difficulty will go wrong," was soon illustrated when the executive secretary became pregnant.

   C. Enclose words used in unusual senses if the reader is not expected to understand them.

   Each of these "bits" of information becomes a small magnetic field on a memory tape within the computer. (Note that once such a term is established in the field or even defined within a paper or article, the quotation marks are dropped.)

7. The *question mark* should be used to:
   A. Indicate a direct question (but not an indirect question).

   Thus, near the middle of the election campaign, each member of the panel was asked two questions: (1) "Have you tried to convince anyone of your political ideas recently?" and (2) "Has anyone asked your advice on a political question recently?"[4]

B. Indicate the writer's uncertainty about the preceding word, figure, or date.

Edward Taylor (1642?-1729) is a prime example of a Puritan minister whose outward life concealed an esthetic response to his theology that was at once both Puritan and un-Puritan.

8. *Parentheses* should be used to:
   A. Set off parenthetical, supplementary, and illustrative material.

   Some of the possibilities of making intelligent use of "noise" in an artefact designed to take account of both structural and metrical aspects of information (in the semantic sense) have already been outlined by the author. (Paper hereafter referred to as CDA.)[5]

9. The *bracket* should be used to:
   A. Set off editorial corrections in a quotation.

   "The total number killed was 484 [Coltrin says 473] of which all but one died of asphyxiation."

   B. Set off editorial interpolations in a quotation (use parentheses within brackets).

   "In regard to his [Maeterkume's] theory...."

It is not felt necessary to include here a discussion on the use of the period or the exclamation point. Regarding the latter, however, a word of caution: in technical writing the exclamation point is seldom used. If you find yourself using it, you should check to see if there is not a better way to state your idea.

—The Paragraph

What is a *paragraph?* A paragraph is a sentence or a group of sentences developing a single topic.

As a functional unit in writing, its strength lies in its unity, coherence, emphasis, and proportion. Consequently, paragraphs have no set length, although length must always be a consideration. Too often, long and involved paragraphs are difficult to grasp, and rewriting the paragraph into smaller thoughts can make them easier to understand. Usually, logical analysis determines the length of the paragraph, but the need for "eye relief" on the page is a second and subtle criterion in paragraph division.

The paragraph is, of course, a unit of the whole paper, and it contains a predicted amount of the material of the paper, usually the amount of material covered by a specific entry in the outline for the paper. Thus, in a brief paper of, say, three pages and having three main topics, with two to four subtopics for each, the paragraph would contain the material in a subtopic. For a larger paper of ten to twenty pages, a paragraph would probably correspond to a sub-subtopic.

BRIEF PAPER

Introduction   One paragraph introducing entire paper
  I. One paragraph (or a sentence) introducing Topic I
     A. One paragraph
     B. One paragraph
     C. One paragraph

II. One paragraph (or a sentence) introducing Topic II
    A. One paragraph
    B. One paragraph
III. One paragraph (or a sentence) introducing Topic III
    A. One paragraph
    B. One paragraph
    C. One paragraph
    D. One paragraph
Conclusion   One paragraph concluding entire paper.

Correlating paragraphs with the outline in this fashion is one of the best means of insuring that paragraphs will be unified and coherent.

In technical writing, each paragraph should have a topic sentence, stated or implied. The topic sentence expresses the central idea of the paragraph. Although a topic sentence may come anywhere within a paragraph, the central idea of a technical report paragraph is usually stated in the first sentence.

In the following paragraphs, the topic sentence is italicized:

> *The American newspaper was making great strides by the mid-nineties both as an instrument of society and as a business institution.* Thanks to the publishing concepts and the techniques of the new journalism, many dailies had won larger circulations giving them greater public influence and support. Their increased advertising and subscription revenues provided the resources necessary for more intensive coverage of the news. And growing financial stability meant that conscientious publishers and editors could do a better job of telling the news honestly and fully, and of demonstrating their community leadership, because they were better able to resist outside pressures.[6]

There is a striking relationship between density and knowing one's neighbors by name. Particularly in the more dense neighborhoods, the closer one lives to neighbors, the fewer are known by name. In fact, only eight percent of the residents living in the densest neighborhoods felt they knew all the adults in the "half dozen families" living nearest to them by name, compared to almost two-thirds of the residents in the least dense neighborhoods.[7]

—Summary

Ultimately sentences and paragraphs form a logical, discernible structure within the paper. In that structure, the sentences are the bricks, the paragraphs are the walls, and the walls are laid out according to the plan of the outline. Viewed this way, the writing of sentences and paragraphs, and the providing of word or punctuation signals within them, is just the process of orderly layout of information so that the reader can readily understand it.

Chapter 4

Science is built with facts
as a house is built with stones,
but a collection of facts
is no more science than
a heap of stones is a house.

—Jules Henri Poincare

Defining the Problem

All too often the writer's most difficult task is that which faces him first—adequately defining his problem. Inadequate delineation of the scope of the paper results in gathering material not pertinent to the problem and in organization that does not lead the reader to logical and convincing conclusions.

On the other hand, if the writer really understands the problem he is writing about, be it a class report, a term paper, a thesis, or a professional article, he has, as a result of that understanding, the direction necessary for efficient gathering of information.

So, before beginning the report or before proceeding very far with the research for the report, the researcher should ask himself what he is about. The definition of that task usually involves these six preliminary procedures.

—*Procedures in Defining the Problem*

1. Make an overview of the problem.
2. Narrow the subject to workable size.
3. Review the literature.
4. Define the independent and dependent variables.
5. Construct a hypothesis.
6. Prepare a preliminary proposal.

—*Overview of Problem*

An excellent way to approach the definition of a problem is to place it in a broad overview. This placement can be handled in three easy steps.

Step one would be to consult those sources which give general statements on the subject. Sources to consult could include first, an encyclopedia, and second, a basic text on the subject.

A general encyclopedia gives the broadest possible view of the subject. It provides terminology, definitions, and the relationship of the topic to other fields. It will often provide also a basic bibliography and the names of eminent authorities on the subject.

Following the general encyclopedia review, a review of a basic text will provide a deeper and more comprehensive view of the subject. The reading of that basic text will, of course, be easier and more productive because of the general background provided by the encyclopedia. Like the general encyclopedia, the basic text will often provide names of authorities and additional bibliography.

The rest of the research is just a continuation of this basic process, but at each step the research and the search for knowledge gets deeper and narrower.

Step two would be to consult a sampling of sources which are semispecialized in their approach to the subject. Sources to consult here could include advanced and intermediate texts and monographs and series published on the subject.

Step three would be to consult those sources which give highly specialized information on the subject. Sources to explore in this stage of the investigation include all the primary documents in the field. (But by this time it is a rather narrow field.) Only if the student completes, in order, the suggestions offered in step one and step two can he be certain he is ready to put into proper perspective the detail offered him in the specialized journal articles, reported field research, and other material which comes to him in this final step.

(Many will have recognized a very effective, though probably incomplete, review of literature in the foregoing three steps. The next chapter suggests the way the information surveyed should be recorded.)

Now, with the background necessary to place his problem in social theory, the investigator is ready to define his problem. (If the paper to be written is the product of major research, the final definition of the problem will take the form of a research proposal.)

—*Narrowing the Problem*

A freshman student proposed to write a paper with the title "The Pacific Ocean." While the subject was both too broad and too deep for him, his temerity in attempting to treat it in the assigned ten pages is familiar to anyone who has supervised research at the undergraduate, graduate, or even the professional level.

A common theme in social science methodology classes is (1) narrow the problem, (2) narrow the problem, (3) narrow the problem. But how to "narrow the problem" still requires some tools. Obviously, if the subject is an applied one, the problem narrowing is done by the research supervisor, and the researcher then writes on the portion of the subject allotted to him. But in the academic world, whether for a class paper, a thesis, or a piece of professional research, the topic must be cut down to size by the researcher himself. The following procedure is one that works well.

Limit by time. This is the most obvious way of cutting down a subject. History courses cover a subject by period, either by natural division such as the period between two important wars (1865-1918) or by arbitrary divisions (1850-1900). Similarly, time limitations can often be used in other disciplines to cut a topic down to size.

Limit by geography. The space limitation can also be applied either with the kind of time limit just described or by itself. The limits can be by political subdivisions (nation, state, county, city, ward, block) or by physical geographical limitations (continent, climatic region, river, drainage, valley, etc.).

Limit by discipline. The body of academic knowledge in the world is, in the university, divided into colleges and departments. Taking only the facet of a subject that would be the concern of a college or department lets the researcher look at his subject more closely. Again, this technique can be used alone or in conjunction with the previous two. Thus a subject like "Demographic factors in the election of 1896 in Humboldt County, Nevada" has been limited by time, geography, and discipline.

Other limiting devices. A social science subject can be limited by such factors as sex, age group, political party, social or economic level, etc. All these limiting factors not only hold the topic to appropriate size, but also may force the researcher to look at his topic more logically and more thoroughly than he might otherwise do. They help him avoid the pitfalls of sloppy generalization, and ultimately, and surprisingly, let him say more about a small subject than he could say about a large one.

—Review of Literature

The purpose of the literature review is to establish the relation of the investigator's research to the larger field. This is done by summarizing the theory, methodology, and findings of research conducted in the area under study.

> The literature summary, then, should include integrated general statements, each statement containing as much information as possible. If one finds, say, five studies on a particular subtopic in the area, he might try to formulate a single sentence which will summarize them all. In this way, the major results in a given field can be surveyed in no more than three or four pages. The more condensed and brief the summary, the more perspective it provides and the more apparent the inconsistencies become.[1]

On major research projects, the student's major advisor will determine the scope of the literature review. Some advisors will require a separate section or chapter labeled as such. Within this separate handling, some advisors will require a detailed, and hence, long review; others will prefer a shortened or condensed version. Indeed, what seems to be a growing number of advisors feel that the only literature review needed is the theoretical base of the study.

We will not be dogmatic about which of these positions is best. Obviously, a researcher must satisfy his advisor or his editor, and that advisor or editor is in the position to know or dictate what kind of literature review best suits the circumstance. For this reason it is imperative that the researcher confer with him about what the expected review should include.

—*Independent and Dependent Variables*

In the social sciences, a common approach to specific problem definition is to identify the independent and dependent variables of the study. In categorizing the variables of a study into this dichotomy, the independent variable is the antecedent of the action and the dependent variable is the consequent of the action.

In an experiment, the independent variable is the variable manipulated by the experimenter. (If the study is the type where the investigator cannot manipulate any variables, that is, nonexperimental research, the independent variable can still be presumed to have been manipulated before the investigator got the problem.[2]) For instance, if one were to study the mass media and rumor construction in a population, the availability of the mass media could be ordered as the independent variable (cause) and the amount of rumor construction could be ordered as the dependent variable (effect). The dependent variable, then, is the condition the investigator is trying to explain (the amount of rumor construction taking place). On the other hand, the independent variable is that which is predicted *from* (in this case, the availability of the mass media).

As changes in the independent variable take place there are resulting changes in the dependent variable.

As an exercise, the student is encouraged to identify these two variables in reported studies. Further uneasiness in handling the independent and dependent variables can be overcome with repeated exercises of this kind.

—Hypotheses

After sufficiently narrowing the scope of the problem the result is the definition of the problem—a thesis statement. This statement is usually abstract. It is put into its specifics by the formulation of hypotheses, or testable statements.

> A hypothesis is a conjectural statement, a tentative proposition, about the relation between two

or more observed (sometimes unobservable,
especially in psychology and education) phenomena
or variables. Our scientists will say, "If
such-and-such occurs, then so-and-so results."[3]

Although a hypothesis comes from a question asked
of nature, it is conventionally written as a declarative
sentence. Further, it is written as an expression of the
expected outcome. Each good hypothesis fulfills two
criteria: (1) as noted above, it contains a statement about
the relationship between variables, and (2) it contains
a statement about how this relationship is to be measured
(if applicable).

The following is an example of a hypothesis that
fulfills both of these criteria.

> Rumor systems in communities served by a
> home-produced daily newspaper will be smaller in
> size—as measured by the number of contacts made
> in rumor transmission—than rumor systems in
> communities not served by a home-produced
> daily newspaper.

A second form for expressing a tentative assump-
tion is to write a null hypothesis—a statement of no difference, or no relationship, between two variables. The
previous hypothesis rewritten as a null hypothesis would
say:

> Rumor systems in communities served by a
> home-produced daily newspaper will not be
> substantially different—as measured by the number
> of contacts made in rumor transmission—from
> rumor systems in communities not served by a
> home-produced daily newspaper.

The null hypothesis lends itself particularly well to testing and statistical analysis.

Most hypotheses, including the null hypothesis, can be reduced to one (sometimes lengthy) sentence.

Without going into the rationale as to why, the scientist cannot prove a hypothesis. This has a bearing, then, upon the wording used in a report if, indeed, the investigator did "prove" something. Kerlinger suggests wording such as "The weight of evidence is on the side of the hypothesis," or "The weight of evidence casts doubt on the hypothesis."[4] Braithwaite says that in "suitable cases we may say that it establishes the hypothesis, meaning by this that the evidence makes it reasonable to accept the hypothesis; . . ."[5] Scientists also say the evidence (or data) supports (substantiates, sustains, etc.) the hypothesis, that the statistic is in the direction of (bears out, corroborates, etc.) the hypothesis, and the like.

One final note about a hypothesis. While it brings to social science research a high degree of precision in handling the data, a hypothesis is not mandatory to all social science research. Many excellent studies are completed each year, particularly in history, economics, and geography, that have not utilized hypotheses. Even the behavioral scientist, as Young notes, sometimes proceeds without a hypothesis.

> It should not be assumed that a study must of necessity proceed from an hypothesis. Many scientific studies were begun and successfully carried forward without any particular theory to prove or disprove. Robert S. and Helen M. Lynd maintained that they made no attempt to substantiate a thesis when contemplating or when writing their final report of the famous study of Middletown. Generally monographic studies such as *Middletown* aim to study and record social data as comprehensively as possible, exploring each

significant and related fact as it is uncovered.
Absence of a working hypothesis does not mean,
however, absence of definitely formulated objectives,
or of basic assumptions upon which the study is
built. The Lynds had their assumptions and
objectives carefully outlined and thus
had a starting point.[6]

—*Preliminary Research Proposal*

Major research dictates the need for a preliminary
research proposal. The proposal is simply a statement on
how one proposes to do the research. For a thesis or
dissertation, the proposal is usually about a half-dozen
pages in length. But it can be one or two pages or one
or two dozen pages in length. Some large research projects for the federal government may even have proposals
several volumes in length.

The preliminary research proposal can serve other
functions to the writer beyond that of satisfying an advisor or committee that the writer knows what he is doing.
The detailing of the research on paper very often points
out to the investigator himself errors in both his theory
and design. Such an awareness can save the investigator many problems at a later stage in the investigation.
In some cases, it may even point to the fact that the
problem—at least in its present form—is not really
researchable.

The proposal should adequately cover four major
points:

1. It should clearly state the problem to be
   studied, or the central theme of the study.
2 . It should place the study in social theory.
3. It should set forth the methodology and the
   types of analyses that will be employed.
4. It should give anticipated results.

So long as the writer covers the above points, the particular form he uses is inconsequential. Therefore, the following outline of a proposal is only suggestive.

—Proposal Outline

Statement of the Problem. This part of the proposal should set forth very clearly to the reader the intent of the investigation. (Note the relationship of this to *purpose* as discussed in Chapter II.)

Place in Social Theory. This section should establish the investigation's relationship to the greater whole of the area in which the research falls. If desired, the review of literature, in a condensed version, can also be included.

Results Anticipated. This part of the proposal lists the anticipated results of the study, often in the form of hypotheses (in which case the heading would preferably read "Hypotheses"). If there is theory building in the form of hypotheses, it should also be handled here rather than in the social theory section. Alternative hypotheses or results should also be discussed in this section.

Justification of Study. The research gaps that the study proposes to fill can be discussed in this section, as well as the study's relation to such other research factors as time limitations, financial considerations, generalizations of anticipated results, etc. Because of its content, the justification of study is often omitted.

Methodology of Study. This should be a fairly detailed report of how the hypotheses are to be tested. It should tell what, how, and why samples will be selected, the manner of manipulation of the independent variable(s), descriptions of instruments used, pretests that will be conducted—in short, a full review of the data gathering procedures. This section might also include a listing of personnel, their training, and the research time schedule.

Types of Analyses. The preliminary proposal also calls for a statement regarding the data analysis methods to be employed. It is here that such aspects as coding, statistics, significance tests, and other analysis methods are outlined and justified.

Summary. Desired concluding statements are reserved for this section. Often the study justification and the summary are combined.

Bibliography. If desired or requested.[7]

—Summary

Preparing to write a paper, thesis, or article in the social sciences obviously involves much more than starting to write. Besides gathering the information, it involves gaining an understanding of how the subject of the paper fits into the total subject or field. It involves a careful defining of independent and dependent variables. It involves a careful narrowing of the problem to a size that can be handled—a procedure that may have to be repeated several times. After that, it involves making a carefully stated hypothesis and may involve incorporating that hypothesis into a formal research proposal. Following the proposal (or perhaps a part of the proposal) is a search of the literature to see what has already been done.

All of these preliminary steps are needed if research is to be efficient and thorough. Failure to follow them usually results in wasted motion and research that is not trustworthy.

Chapter 5

Read not to contradict and
confute; nor to believe and take
for granted; nor to find talk and
discourse; but to weigh
and consider.

—Francis Bacon

Gathering
Information

Once the problem has been defined, the writer is prepared to begin the complex process of gathering information. This process can be long and tedious. It can also be rewarding and exciting. The purpose of this chapter is to acquaint the researcher with the major sources of information available to him in a literary search.

Often the suggested routes to gathering information seem overly detailed to a beginning writer. However, time spent now with detail will be time saved at later stages in the preparation of the report.

Initially, in formulating the approach to gathering material, the writer should make certain he is using all the available sources. Generally, the sources fall into three categories: first, publications in the area (usually found at the library); second, other people working in the area; third, the writer's own work.

—*Publications*

The complaint is too often heard that the social scientist has no idea of the wealth of information that awaits him in the books, journals, monographs, and other material currently in print. Since utilization of previous knowledge is the basis of progress, the success of the social scientist depends to a great extent on his ability to make use of the knowledge that others have uncovered.

The task at hand, then, is to know how to locate this published material. Incredible amounts of information are available in good libraries, and that information, in catalogs, references, abstracts, indexes, and yearbooks published throughout the world, is carefully catalogued and indexed ready and waiting for the able researcher. All that is necessary is to know the techniques of finding it.

Many libraries—particularly college libraries—have a general reference complex which serves as an information center. The writer who does not know where to begin a literary search should request assistance at this center. Not only will librarians answer questions on library problems, but in many general reference centers they will also teach the use of the card catalog and other bibliographic skills basic to beginning research. If the library has such instructional programs for their own holdings, it is generally wise to take advantage of them. In the absence of such special programs tailored to a specific library, the following library research procedures will work for any library.

—*Card Catalog*

Almost all the materials in a library are organized as if their sum and total constituted an enormous book in which each body of subject material is a chapter, each volume a page. The index to this giant book is the card catalog. Contained within the card catalog are cards (arranged in alphabetical order) providing access to all classified information in the library by author, title, or subject.

Author-Title Catalog. Use the author-title catalog when you want a specific work and you know its author or title.

Authors are usually persons, but they may also be:

Associations: e.g., National Education Association of the United States.

Societies: e.g., American Catholic Sociological Society.

Government Agencies: e.g., United States Department of Commerce.

Cards are arranged alphabetically word by word by the top line of the card, and the shorter word is filed before the longer word that begins with the same letters.

New wives for old

Newark; a poem

Newman, Alfred E.

Newmann, Kurt

Cards for the same author are arranged alphabetically by the titles of the books.

Many books are also listed under their titles; however, you will seldom find titles with commonplace beginnings like:

History of . . .,

Outline of . . .,

A title beginning with an initial comes before a word commencing with the same letter (ALA Bulletin, Alabama highways). A title containing the name of the issuing body is often filed by that name (*American Medical Association Journal,* not *Journal of the American Medical Association*).

Sample of Periodical Card

```
Ref 1
301. 1505
B394      Behavior today; the human sciences
               newsletter.    Del Mar, California.




               I. The human sciences newsletter.
```

Subject Catalog. In the subject catalog you can find books on a subject without having to know the names of authors or titles, since the cards are arranged alphabetically by subject. The subject is typed in red at the top of the card. Be specific in looking for a subject. For example: Look under France—History, not History.

Some subjects are divided and subdivided

    by geographical areas: Education—Arizona

    by periods of time: Great Britain—History—
       1760-1789

    by the form of the publication: Sociology—
       Dictionaries

For books *about* a person, look under his name in the subject catalog; for books *by* him, go to the author-title catalog.

Subject Cross References. In the subject catalog are so-called "see" cards which direct you from a subject term not used in the catalog to the term that is used. For example:

> Aviation
>
> see
>
> Aeronautics

In other words, all books on this topic are entered under "Aeronautics," not under "Aviation."

To find related aspects of a subject, refer to the "see also" (sa) references listed under your subject in the *Library of Congress Subject Headings.* Then consult the subject catalog under these related headings.

In the subject catalog, when there is more than one card representing a topic, the cards are arranged alphabetically by the authors' names.

Sample of Subject Card

```
                  Arbitration, International
341.52
B155p     Bailey, Sydney Dawson.
                Peaceful settlement of disputes: ideas and
          proposals for research, (by) Sydney D. Bailey ...
          (New York) United Nations Institute for Training
          and Research (1970)
                2 p.l., iii, 67p. 23cm.
                Reproduced from typewritten copy.
                1.Arbitration, International. 2.International
          relations.   I.United Nations Institute for
          Training and Research.  II.Title
          MH-L                          NUC70-99190
```

Sample of Author Card

```
341.52
B155p    Bailey, Sydney Dawson
              Peaceful settlement of disputes: ideas and
         proposals for research, (by) Sydney D. Bailey ...
         (New York) United Nations Institute for Training
         and Research (1970)
              2 p.l., iii, 67p.  23 cm.
              Reproduced from typewritten copy.
              1.Arbitration, International.  2.International
         relations.  I.United Nations Institute for
         Training and Research.  II.Title.
         MH-L                              NUC70-99190
```

Sample of Title Card

```
              Peaceful settlement of disputes.

341.52
B155p    Bailey, Sydney Dawson.
              Peaceful settlement of disputes: ideas and
         proposals for research, (by) Sydney D. Bailey ...
         (New York) United Nations Institute for Training
         and Research (1970)
              2 p.l., iii, 67p.  23cm.
              Reproduced from typewritten copy.
              1.Arbitration, International.  2.International
         relations.  I.United Nations Institute for
         Training and Research.  II.Title.
         MH-L                              NUC70-99190
```

Note that this card is the same as the one above except that the title has been typed in at the top. The card is then filed in the "p" part of the alphabet.

Keep the following cataloging rules in mind:

1. The articles "a," "an," and "the" and their foreign language equivalents are disregarded in filing when they are the first word of a title (except for proper names, e.g., Los Angeles).

2. Abbreviations are arranged as if spelled out in full.

   Dr. as Doctor

   St. Louis as Saint Louis

   U.S. as United States

3. M, Mac, and Mc are filed together as if all were spelled Mac and interfiled with words spelled similarly.

McAdam	Machinery
MacArthur	McVey
McHenry	

Classification. Most books in most libraries are classified and arranged by number so that books on a subject will be shelved together and related books will be near at hand (a great help to the researcher in open-stack libraries). The most common numbering or classification system used is the Dewey Decimal system. In this system, a basic classification number is expanded by means of the decimal to represent further subdivision, and a letter symbol is added to designate the author of the work. The classification and the author numbers together make up the *call number,* the coded symbol by which a book may be located in the library. In addition, some books carry *special call numbers* because they are housed in special collections or elsewhere than in the main stacks of the library. Most information central to the social sciences is found in the 300 and 900 classifications.

Because they wish to accommodate many books under the same classification number without going into involved decimals, some libraries (usually the very large) use the Library of Congress Classification, which bases its system on the alphabet. Beyond a general number from A to Z, each of the subject divisions is further divided by letters and numbers to show the exact classification of each book. Under this system most information central to the social sciences is found in the C through J classifications.

For both systems, the classifications most useful to social scientists are listed below:

Dewey Classification

020-029	Behavioral Science
060-069	Behavioral Science
130-139	Behavioral Science
150-159	Behavioral Science
300-359	Social Science
360-379	Behavioral Science
380-389	Social Science
390-399	Behavioral Science
570-573	Behavioral Science
640-649	Behavioral Science
650-659	Social Science
900-999	Social Science

Library of Congress
Classification

AM AS	Behavioral Science
BF	Behavioral Science

C D E F G	Social Science
GF-GT	Behavioral Science
H-HJ	Social Science
HM-HV	Behavioral Science
HX J K	Social Science
L-LT	Behavioral Science
TX	Behavioral Science
U V	Social Science
Z4-Z1039	Behavioral Science

References. In addition to the card catalog, there are many reference publications that serve as excellent guides in the literary search. For all social scientists, the following are very useful.

Winchell, Constance M. *Guide to Reference Books.* Chicago: American Library Association, 1967.

"Winchell" is the standard classified and annotated bibliographic reference guide for all fields. The index includes author and subject entries and most title entries.

White, Carl M. *Sources of Information in the Social Sciences.* Totowa, New Jersey: Bedminster Press, 1964.

This extensive guide to the literature of the social sciences treats, respectively: social science in general; history; economics and business; sociology; anthropology; psychology; education and political science. Each chapter includes a bibliography eassy on the discipline prepared by a specialist and an annotated list of reference sources and periodicals. There is an author and title index, but no subject index.

Supplementing the above in a literary search are indexes and abstracts. Indexes list books, articles, reports, proceedings, yearbooks, and other materials according to some system, e.g., by author, title, subject, form, date of publication. They are of two types: general, such as *Reader's Guide to Periodical Literature*; and specialized, such as *Education Index*. Abstracts list published sources according to some system, but in addition they summarize the essential features and facts of the source. Without indexes or abstracts to the millions of periodical articles published each year, the research worker would be hopelessly lost.

The usefulness of any reference work depends on you, the researcher, whose particular needs and situation will be the ultimate determining factors. Consequently, you should take the time to familiarize yourself with the major references in your field. To do this, you should seek the assistance of the social science librarian where you do the bulk of your research. Very often such librarians will already have prepared a guide to the literature in that library.

Magazine Articles. Perhaps the most current information is to be found in magazine articles in such publications as *Harper's, Atlantic Monthly* and the *Saturday Review*. The best way to find magazine articles on certain subjects is through general and specialized periodical indexes available in libraries. The most general index is the *Reader's Guide to Periodical Literature,* noted earlier. Here is an entry as it would appear in the *Reader's Guide*:

> GEOGRAPHY
> Mystery island: a lesson in inquiry. J. Zevin
> il Todays Ed 58:42-3 My '69

This tells the reader an article on the subject of geography entitled "Mystery Island: A Lesson in Inquiry" was written by Jack Zevin. Further, it contains illustrations, it appeared in *Today's Education,* in volume 58, on pages 42 and 43 (43+ would have meant that the article was continued on other pages), and it was in the May 1969 issue.

This information will also appear under the author's name elsewhere in the guide.

Other useful guides in the social sciences include the *New York Times Index, Public Affairs Information Service Bulletin, Monthly Catalog of United States Government Publications,* and *International Index to Periodicals.*

Newspaper articles. Four excellent indexes to newspapers are the *New York Times Index, Times* (London) *Index, Wall Street Journal,* and the *Christian Science Monitor Index.* If, for example, you need newspaper articles on labor strikes, you would look under that topic in, say, the *New York Times Index,* which is an alphabetical index to the subjects, persons, and organizations covered in articles appearing in the *New York Times.* Under each subject the arrangement is chronological.

A full page of instructions on "How to Use the *New York Times Index"* appears in the front of each volume.

Although the *New York Times Index* is a guide to the contents of that newspaper, it can also be helpful in locating articles in other newspapers because it gives the date of a national or international event.

Human Relations Area Files. Many social scientists are fortunate enough to have access to a library that houses a Human Relations Area Microfile. An outgrowth of the Cross-cultural Survey begun at Yale University, the microfile is an extensively indexed and annotated collection of source material on the culture, behavior, and background of a representative sample of primitive, historical, and contemporary peoples and societies around the world. In these topically indexed microfiles are some eight hundred categories of data.

Government Documents. In many libraries the government publications which the library receives are not listed in the card catalog. The keys to these collections are printed bibliographies and indexes such as the "Monthly Catalog of United States Government Publications" and the *United Nations Documents Index.* Ask the librarian for assistance in finding what you need.

Maps. Ask the librarian to help you find the map you want.

Theses. These are usually listed in the main card catalog just as books are, i.e., author, title, subject. They are also listed by departments in some libraries. The titles carried, however, are most often local in nature. Theses and dissertations do not usually appear in the regular book trade bibliographies, but because they are contributions to knowledge, they do have a definite place in the research plan of the social scientist. For this reason, the student should check with the local librarian on which of the various publications your library does carry that gives information concerning this research output.

Interlibrary Loans. Another excellent source of material is the interlibrary loan service, by which the library procures from other agencies materials not held in its own collections. Application forms can be obtained and filled out in the subject library where you would expect to find the title if the library had it. Be prepared to supply full bibliographical data for the volume to be borrowed.

Finally, if the library does not have what you need, there are some other alternatives open to you. If your need is not immediate, you may ask the library to buy the material. Often a thesis advisor or class instructor is willing to have the library order books and other materials.

Card File. The heart of the research program is the bibliography, an index of the source material available to the writer. Experience has proven the worth of the card system as a means of recording both the bibliographical reference and the notes. It is not, as students sometimes assume, a valueless busywork procedure, but a sound method of research that saves much more time than it takes.

Use a separate three-by-five-inch card for each bibliographical reference. This will constitute the working bibliography. When completed, the working bibliography is the final bibliography of your paper; all that remains is to put the cards in alphabetical order and type off the bibliography. (Remember: the card file will accumulate references that will prove worthless; periodically these should be removed but saved, since the same titles may be met again, and the researcher cannot always remember that he has previously consulted them.)

On each bibliography card should be certain basic information. At the outset cultivate the habit of including all this information; missing information will require a second—and needless—visit to the library (a publication date discovered to be missing at midnight before the paper is due must remain missing). This basic information includes: (1) author's full name, (2) title of the reference, (3) facts of publication: for a book, the place of publication, publisher, and date of publication; for a magazine, the name of the publication, volume number, date of issue, and page reference; for a newspaper article, the name of the newspaper, date of issue, page reference, and column number.

In addition, as optional information, you may want

to include the call number of the reference, personal assessment of the reference, and other such notations.

When recording the reference, the entry should be in the same style as will be used in the paper itself.

Note File. Use a four-by-six or five-by-eight-inch card for note taking. These sizes of cards provide the space needed for adequate notes, and also readily distinguish a note card from a file card. Just as using file cards permits easy arrangement for the final bibliography, the use of note cards permits the easy arranging of material in the order the student wishes when typing up the first draft of the paper.

To aid later organization use separate cards for each kind of information. Rarely will more than one card be needed for a properly digested note.

Give each note card a useful topic heading. Such titles come from a constantly-revised working outline.

Write the source in the upper right corner of the note card. It is not necessary to repeat all bibliographic information on the note card, but indicate the source in some consistent place either by a key number of the bibliography card or by a short title of the source. Be sure to include the page number.

The notes themselves may be either *exact* quotations or paraphrases of the original. Be sure to indicate which it is. We have met numerous cases of apparent plagiarism which ultimately resulted not from dishonest intent but from sloppy note-taking procedures.

For a quotation, take down exactly what the author says. If you want to follow this with your own comments which will be helpful to you later when you can no longer see the note in context, put those notes in square brackets, [ ], to distinguish your own text from that of the author you are quoting.

If you paraphrase the original, be sure that you change the structure substantially rather than just supplying occasional synonyms for the author's words. Further, you should label your paraphrase as a paraphrase in your notes.

Incidentally, you should not take notes in pencil since it smudges and become illegible as the cards are rearranged. Retyping of written note cards is a waste of time. Write with pen, and write legibly enough so that you or your typist can read accurately what you have written.

Philosophy of Documentation. Regardless of whether you quote or paraphrase material in your notes, you will need to document all material that is identifiable as having come from a particular source. Consider the following situation:

You are to read your paper before a meeting of an academic society. Present are the eminent experts in the field—including the authors of your source books. Copies of your paper are distributed to the audience. Now, what would you document?

Obviously you will want to give full credit for any exact words that you quote. You will also want to give credit for a specific theory, line of reasoning, or set of statistics that you cite, even though you do not use the exact words of the original author.

You would, of course, avoid giving credit to Schwartz for a line of reasoning that he mentions but has drawn from the earlier work of Maetterklume. (Maetterklume may be sitting in the fifth row.) You would also avoid giving credit for any piece of standard information that is widely available even though that information may be rather specific (for example, the speed of light is 186,300 miles per second). An even more specific value resulting from a recent study using lasers might be cited specifically for source, but the 186,300 figure can be regarded as commonly held and accepted knowledge.

Failure to document properly in published works may make the writer liable to civil and criminal action under copyright laws. But even if the paper is not published, improper documentation is a discourtesy to the original researcher. At least in the academic world, it may in serious cases be regarded as evidence of dishonesty and basis for academic discipline or expulsion.

—*Other People as an Information Source*

Researchers, especially student researchers, often neglect to take full advantage of the knowledge and experience of other social scientists. Yet some students who have tried to interview researchers have complained that they did not get good cooperation. Our own experience is that the interview can be extremely effective— far more effective than the questionnaire for in-depth material. But the interview, more than any other research technique, requires tact and consideration. Most ineffective interviews have failed because of the lack of such courtesy. The following guidelines will prove helpful.

1. Make an advance appointment. (Such an appointment can be made by phone. For an interview with a professor, it should be set during a regular office hour, and such hectic times as registration and final exam times should be avoided. The interviewer should, of course, arrive on time.)

2. The interviewer must be prepared and indicate that he is prepared. The interviewer who arrives and says, "I am doing a paper on

population control. Tell me all about it," will nearly always get a brush-off.

A well prepared student might say, "I am doing a senior paper for Professor Willoughby's class on population control in India in the last decade. I understand that you have done some work on the subject. I have read the works of Jones, Schwartz, and Maetterklume, which have been helpful, but I have not been able to obtain copies of current issues of *Indian Demography*. Can you suggest where I might find them or find other material pertinent to the subject?"

Students who have used this approach have usually found professors and other researchers more than willing to assist, because the student has shown his willingness and capability for work on his own.

3. Careful notes should, of course, be kept. The date and the place of interview, the full name and title of the interviewee, the exact questions asked, etc., should all be a matter of record. A tape recorder may be very handy for such an interview, but you should always ask permission before making a recording of a conversation, or a lecture, for that matter.

4. Remember throughout that you are there to listen, not to instruct or debate. Talk no more than is necessary to keep the interviewee on the subject.

5. Keep the interview brief. (Prepared questions will help here.) When you have finished, thank the interviewee and leave. Of course, if the man wants to talk, let him. Such talk

from a man who is an expert on a subject can be a rare educational experience, but don't mistake mere courtesy for an invitation to stay all day.

6. Send a thank-you note afterwards and make appropriate acknowledgment in your paper for help received. This is common courtesy, and it makes the way smoother for subsequent students—yourself included.

Most of what has been said about gathering information from other people has referred to other researchers, but the same basic principles apply to getting information from eyewitnesses to events or to getting information or opinions from people who are the subject of the study. In all, we cannot overemphasize the need for courtesy, tact, and delicacy to avoid inconvenience and unnecessary invasion of the privacy of another person. The middle-aged and elderly may be easily offended by brashness—a fact that youth may easily forget.

—*Personal Research as an Information Source*

For anyone but the very rare individual with a photographic mind it is necessary to keep notes. Those notes, if properly kept and filed, may become an important source of information in later months and years.

The notes that are taken as part of a specific study are usually fragmentary, but the broader, though perhaps shallower, notes that can be kept as a write-up of all your reading may be helpful in ways undreamed of when the original study was done. The following suggestion of reading procedure works well.

1. Always read the preface and introduction to books and the abstracts that precede journal articles. This will make the publication more

meaningful, for such sections offer overviews, approaches, and summaries of the material to be covered.

For instance, notice how Lewis Mayers, in his preface, establishes with his reader his approach to material presented in *The American Legal System.*

> Nowhere is there to be found, between two covers, a systematic account of our legal institutions—as distinguished from our law—in all their varied aspects. This lack the present volume purposes to supply.
>
> The account is not restricted to an exposition of the current situation. The historical roots of the chief legal institutions of today are also traced. And, too, current proposals for reform, to which the author has made bold in some cases to add proposals of his own, . . .
>
> The volume is not confined to the field occupied by the courts. The vast area of adjudication by administrative tribunals is reviewed at some length, as is also adjudication by voluntary arbitral tribunals, and by the several types of military tribunals, including the special types brought into existence by war and occupation.[1]

With this short amount of reading behind you, it is evident that you can know "what the book is about" before you read it—hence, the material is more comprehensible.

2. Imagine you are the author of the publication and in your own words talk through the article to a friend, real or imagined, explaining the major points, the supporting minor points, and the conclusions drawn. (Audience: student.)

3. Let this summary provide the framework for your written abstract.

4. Relate this article to other articles, noting points of integration and differentiation.
5. Record criticisms of the article which you may have.
6. Select out all the words you do not understand and learn them.

This would include using them in a sentence or two to confirm to yourself your mastery of the meaning(s) of the words.

7. Place the article in social theory.

If you complete these seven steps you can be sure that you have made the article "your own." At this point you will feel the exciting beginnings of being a professional in your chosen discipline.

In addition to such notes, your own past papers and studies can become sources of information. Often major papers such as the thesis or senior paper grow out of gaps in theory and research discovered while doing term reports for undergraduate and other courses. Such papers in your major field are a valuable source of information, and you should early form the habit of making—and filing—a carbon copy of all the work you do.

The combining of several research reports done earlier into a unified whole is also an excellent exercise in integrating the knowledge in one's field—an attribute all students should develop.

Journal. Once you embark upon a significant piece of research you should keep a personal journal, for it is an effective way to facilitate the research project.

A "daily diary" of the total research experience, the journal, whose entries are dated and recorded in chronological order, is an excellent way to record ideas, hunches, hypotheses, questions, and conclusions that may come to you in your research work. It is a place where you can well permit yourself to "speculate rather freely and un-critically at certain points . . . and return to re-examine the speculations with a critical eye later on."[2]

For major research work such as the thesis or dissertation, the journal should include all expenses incurred. Such entries are necessary if the research is receiving funds from certain types of sources and as documentation for income tax purposes.

The keeping of a journal insures that the order of the investigation is maintained, that each step is in its proper time relationship. A journal provides a means for a complete review of what has been done before going on to the next stage of the project. It can also serve as a basis for a discussion of the research activity, if one is necessary.

In the major research project, the journal should also serve as a personal history of the data-gathering experience, including such items as problems encountered in training research assistants and in publishing the project. Recording the problems and the ways the investigator met them—and suggestions as to how they could have been met or avoided—can be a helpful guide, not only for future research by the investigator but for others, too.

—Summary

Information, like gold, is where you find it. But, like prospecting, research involves study of the technique of finding the spots more than luck. The careful researcher will use all possible sources in the library, the knowledge of other people, and his own records in order to produce a thorough, professional work.

Chapter **6**

The outline is the blueprint
of the composition.

—Hodges and Whitten

Organizing
the Data:
The Outline

The first step in organizing the data is to write a comprehensive outline that will reflect the purpose of the paper. This chapter will discuss the structure of this outline and review a general social science framework for the outlining of a report.

It is not uncommon to find a scientist, faced with a report and, perhaps, a deadline, wanting to "get on with the writing." If you succumb to this temptation, the writing of the report can develop into a long and laborious experience. To write without an outline is to chance a time-consuming and frustrating project; to proceed with an outline saves time and reduces frustration.

—Preliminary Steps

Before he undertakes the writing of the comprehensive outline, the knowledgeable social scientist usually goes over his data a final time to secure a thorough understanding of the total situation studied. The importance of this final review—particularly in major writing projects—cannot be overstressed. As Pearson states, "The classification of facts, the recognition of their sequence and relative significance is the function of science."[1] A final review of data aids in this function.

Of course the type of study undertaken will determine what the data yield, but as a second step you should ask yourself some central questions during this final review. Young suggests the following:

1. What are the most significant social situations which these facts reveal?
2. What are the outstanding similarities and differences displayed by them?
3. What social processes do these facts reveal?
4. What sequences are manifested by them?
5. What causal relationships are revealed by these situations?
6. What systematic conclusions may be drawn from them?
7. What new hypotheses can be formulated?[2]

At this point, too, the writer should consider the audience for which the report is intended and consider the purpose of the report. (These questions have been dealt with extensively in Chapter II.) We feel that failure to consider audience and purpose is a primary cause of poor writing. Attention to these questions cannot be overemphasized.

For reports of major research, answers to the seven questions about the consequences of data are particularly valuable, but the questions of the intended reader and the purpose of the report will be important to both major and minor projects. Evaluations of the consequences of data and the questions of reader and purpose are necessary preliminaries to constructing the outline.

—Constructing the Outline

It will be recalled that in the previous chapter on defining the problem, you were reminded that if you really understand the problem you are investigating, you have in that understanding the direction necessary for the efficient organization of your report. Now you are ready to supplement that understanding with (1) the additional insights you have gained from the research experience to this point, (2) the comprehension gained from the total review of your data, and (3) the additional direction gained from the consideration of reader and purpose. In sum, you are ready to write your detailed outline.

A word of caution: The outline is flexible. Indeed, any writer should regard it as a tentative plan that he can change whenever there is a need or advantage in changing it.

—Types of Outlines

The most commonly used types are the key-word outline, the sentence outline, and the topic outline.

The purpose and length of the report will very often suggest the type of outline to make. For reports that have only a few divisions, or for reports that are only a few pages in length, it is useful to employ the key-word outline, an outline style in which a word or phrase is used for both major and minor divisions. In this type of outline, the key word recalls to the writer the information he wants to present to his readers. Experience suggests that the key-word outline is better suited to the behavioral science disciplines than the social sciences as a whole.

The following is an example of a key-word outline.

I. Social characteristics
   A. Age
   B. Sex
   C. Occupation
   D. Income

    II. Residence
       A. North
       B. South
       C. East
       D. West

Such a key-word outline is easy to prepare and easy for a writer to follow. It serves as a useful reminder of the relationship between already known pieces of information and as a taxonomic framework, particularly for rather concrete and such obvious ideas as simple existence.

But the key-word outline cannot make clear subtle and fine distinctions, either quantitatively or qualitatively. When such subtle and highly abstract relationships must be shown, the sentence outline is preferable.

In the sentence outline, all divisions of any rank are expressed in a complete sentence. A compromise between the key-word outline and the sentence outline in both form and function is the topic outline, which uses titles, clauses or phrases.

For most subjects in the social sciences the sentence outline is probably best and is the one we would usually recommend. Its strength lies in the fact that to write a heading as a complete thought requires that you eliminate any fuzziness in your thinking, resulting in greater clarity in the outline. Plainly, when you have completed a sentence outline you are a good deal closer to having the report written than you are when you have completed only a key-word or topic outline.

We do not recommend mixing outline types, since such mixing usually means a shift in thought processes or method or completeness of treatment—and the effect might come through into the finished paper. Obviously, the types should not be mixed at any one level of the outline, though there might be justification for having roman numeral headings as sentences and capital letters as topics (or vice versa).

Certainly, if you consider using a mixed outline and your outline is to be seen by a reader or supervisor, you must recognize that many scholars regard a mixed outline as some kind of hermaphrodite to be avoided at all cost.

The following is an example of the sentence outline.

I. Anthropologists tell us much about initiatory rites and ceremonies which accompany passage from one social state to another. These social states include passing from childhood to adulthood, from virginity to marriage, and from life to death.

   A. The purpose of all such rites is twofold. One purpose is to prepare the individual for the new way he should act. The second purpose is to instruct the members of the individual's social community in the new way they must act toward him.

   B. There are three phases in all rites of passage: (1) separation rites, (2) incorporation rites, and (3) transition rites.

      1. Rites of separation are prominent in funeral ceremonies.

      2. Rites of incorporation are prominent in marriage ceremonies.

      3. Rites of transition are prominent in ceremonies at birth.

   C. Often all three rites of passage are used in the same ceremony.

      1. A newborn baby may be separated from its mother or family.

    2. A newborn baby may be incorporated into the family or clan or tribe.

    3. A newborn baby may be transmitted through baptism, or some other such ceremony, from a biological organism to a social being.

II. Differences in rites of passage ceremonies can be noted by contrasting a primitive society with a modern society.

—Problems in Outline Construction

Occasionally a problem will arise in outline construction when you unwittingly change the arrangement on which the outline is formed. In making the outline, be conscious of the arrangement you are using—time, space, or group—so that you maintain a consistent pattern. The following is an example of an inconsistent pattern of headings.

Settlement of the Far West

    I. Period before 1830   [Time]

  II. The Mormons   [Group]

 III. Oregon Territory   [Space]

An example of a single principle of arrangement would be:

Settlement of the Far West

Time	*Group*	*Space*
I. Period before 1830	I. The Mountain Men	I. The Northwest
II. Period from 1830 to 1880	II. The Oregon Settlers	II. California
III. Period from 1880 to 1912	III. The Mormons	III. The Great Basin
	IV. The Forty-niners	

Two other principles of outlining that occasionally prove troublesome are (1) the mixing of coordinate and subordinate headings and (2) the lack of adequate subdivisions for a heading.

One of the prime reasons for using an outline is that it forces you to clarify your own thinking about the subject. This clarity is reflected in the placement of the major and minor topics in proper relationship to one another. This consistency of minor points supporting major points and major points in logical relationship to one another is what gives the paper its continuity. The following is an example of mixing coordinate and subordinate headings, thus destroying the continuity.

Settlement of the Far West
I. Period before 1830
II. Period from 1830 to 1880
III. The Mormons in the Great Basin

An example of a proper arrangement of coordinate and subordinate headings would be:

Settlement of the Far West
I. Period before 1830
II. Period from 1830 to 1880
    A. The Mormons in the the Great Basin
    B. The California Gold Rush

Since the outline is essentially a system of division, each heading level must be divided into at least two parts. If there is a *I*, there must be a *II*; if there is an *A*, there must be a *B*; if there is a *1*, there must be a *2*, and so on until each division has two or more coordinate parts.

Illogical
Settlement of the Far West
I. Period before 1830

(If there is no other period to be discussed, the heading should simply read, "Settlement of the Far West before 1830.")

Logical

Settlement of the Far West
I. Period before 1830
    A. The Early Explorers
    B. The Mountain Men
II. Period from 1830 to 1880
    A. The Period before the Transcontinental Railroad
    B. The Period after the Transcontinental Railroad

—*Framework for Social Science Outlines*

In the social sciences, technical reports are written within a common framework. This framework corresponds closely to that of the research itself. For a report that may run several hundred pages, such as a thesis or dissertation, or for a report that is only a few pages in length, the framework does not change; the difference is in the amount of detail only. Thus all chapters in a thesis or all sections in a term paper can be grouped under three categories:

I. The Problem
II. The Methodology
III. The Results and Conclusions

In reviewing this general approach, Westly notes:

> There is a . . . logic for the journal article's structure: begin with the problem [I.], then present your attack on the problem [II.], then present your results [III.], then discuss them in the light of the problem you have presented, then summarize. In studies employing a behavioral methodology this

order is almost invariably followed, using major
headings such as *Hypothesis, Procedures, Results,
Discussion,* and *Summary.* For historical and
legal investigations, the wording may be different
but the logic is essentially the same. Instead of setting
apart sections of the report as above, often the
writer merely numbers major segments of his report.[3]

This framework, then, with three general sections, is
the standard format for papers in the social sciences. It
is a format that allows for tailoring according to the
specific needs of the technical paper, whether large or
small. Kerlinger provides the following detailed subdivisions that could be ordered within this general format
for the reporting of a major piece of research.

I. The Problem
   A. Introduction to the Problem
   B. Place in Social Theory
   C. Hypotheses
   D. Definitions
   E. Review of Literature
II. The Methodology
   A. The sample and sampling procedures
   B. The testing of hypotheses
   C. Measurement of variables
   D. Methods of statistical analysis
   E. Pretesting
III. The Results and Conclusions
   A. Results
   B. Interpretation
   C. Implications
   D. Conclusions

Note that this is not an outline; it is a topic format from
which a sentence outline will be written.

We would make some last practical suggestions on
outline procedure.

1. Make a one-page outline first.
2. Revise the outline form as needed throughout the whole research procedure.
3. Make the final outline as detailed as possible. Complete it at least to the stage of a heading per paragraph of final report.
4. Since an outline as detailed as the one suggested above (3) is going to be lengthy, at least for a major report, it will be helpful to make it on paper large enough to include the whole outline on a single sheet. We like to use a newsprint roll-end obtained from a printing shop and a marking pen for a large outline. Such an outline may cover the whole wall or floor of the workroom, but it allows seeing all the parts together and allows proportioning of items to their proper importance. It also allows the writer to place his note cards, charts, tables, and illustrations directly on the outline. From this to the first draft is a very simple step.

—*Summary*

The outline is the vehicle that organizes the data into a unified and coherent whole reflecting the paper's purpose. Faulty outlining results in an inadequate report.

Three pitfalls of outlining that should be given special concern are (1) an inconsistent pattern of headings, (2) the mixing of coordinate and subordinate headings, and (3) a heading that stands alone. In the social sciences the outline is usually prepared from a format of three major sections: the problem, the methodology, and the results and conclusions. This standard format is one that allows for tailoring—within the three major sections—to the specific needs of the paper. In writing the detailed outline, you have the choice of three general types of outline styles: key-word outlines, sentence outlines, and topic outlines. All things considered, sentence outlines are probably best suited to most social science topics.

Chapter 7

Writings are useless unless
they are read, and they cannot
be read unless they are readable.

—Theodore Roosevelt

Writing the Paper: Style, Format, and Language

Style in writing refers to two rather different elements. First, it refers to the manner or technique of writing, which includes word choice, sentence structure, length, and techniques of syntax. Second, it refers to format, to the conventions of documentation, and to the manner of handling numbers, tables, charts, and other similar format items. This chapter will discuss both of these aspects of style as they apply to social science writing. The chapter concludes by discussing other systems of communicating that may for some purposes be more effective than word systems.

—Style as Technique

As we have said before in this book, much writing in the social sciences is not well done. This example from a student paper illustrates a number of common problems:

> American scholars are less likely to consider society as a whole, vis-a-vis the individual. They would rather speak of the relationship between the individual and the society, suggesting mediating processes rather than direct interaction. It is not that they deny the reality of society, it is rather that interaction is among persons within a framework of society—individuals and groups organized in time space, sharing many common goals and expectations, but also constantly carving out new objects and expectations through the process of interaction.

The passage is full of difficulties. Does "They would rather speak of relationships . . ." mean that American scholars customarily speak of relationships between the individual and society or that as a matter of preference they speak of the relationship between the individual and society? Does ". . . suggesting mediating processes rather than direct interaction" mean that when American scholars speak, they speak about "mediating processes" or does the fact that they speak suggest mediating processes are going on, or do the scholars suggest that processes be mediated—whatever that might mean—rather than have direct interaction? If such confusion is not enough, the part of the last sentence that follows the dash is very loosely tied to the rest. Upon close analysis, it can be seen that it might be a clause in apposition to the modifying *persons,* but the reader cannot be sure.

Such lack of clarity appears in professional papers, too. Such verbal fumbles will occur much less often, however, if the writer will do three things:

1. Keep his sentences rather short.
2. Use simple words when he can.
3. Be concrete.

Using short sentences does not mean relapsing into the primer style of the Dick and Jane books. But when the reader is wrestling with difficult concepts and relationships, while trying to understand the technical vocabulary, he should not be saddled with the additional burden of twelve-line sentences.

Using simple words does not mean always abandoning precise technical terms for roundabout nontechnical ways of saying the same thing. The technical terms are sometimes irreduceable. But *vis-à-vis* in the cited passage above is intended to mean *rather than* or *in comparison to*. While the writer might protest that it is natural and right for him to talk that way, we would claim that such usage is ostentatious and motivated by something other than the desire to communicate knowledge.

The necessity of communicating information to other professionals frightens many writers. Perhaps as a defense they hide behind pomposity and pretentious diction. The advice of *Journalism Quarterly* on this matter is sound:

> Scholarly writing is a rather formal kind of communication. Often this results in stuffiness. The effort to overcome stuffiness often results in a casualness inappropriate to such a formal report. The solution to this dilemma lies in the best advice for expository writing in general: write simply, directly, and understandably. If your investigation is worth reporting . . . you are assured of the interest of a substantial number of your fellow scholars. Its merit lies in what you did, not in what you say. . . . (How often you are cited, not how widely you are read, is what counts.)[1]

The drive toward extreme formality is evident in the work of many writers today. Such formality, rather than being evidence of ability to deal with the abstract, is too often the result of an insecurity that moves the writers to hide behind an excessive use of area jargon, complex sentence structures, and other smoke screens. One scientist describes this condition this way: "Any ambitious scientist must, in self-protection, prevent his colleagues from discovering that his ideas are simple. So if he can write his publications obscurely and uninterestingly enough, no one will attempt to read them but all will instead genuflect in awe before such erudition."[2]

Again, the best advice for overcoming too much formality in writing is the same as that for too much informality—"write simply, directly, and understandably."

Technical reports are nearly always written in the third person. This grammatically objective point of view theoretically centers the reader's attention on the facts rather than on the investigator. Another aid to gaining this impersonality is the use of the passive voice. This voice is extensively used in describing procedures.

To the inexperienced, however, such procedure causes many problems. Consequently, it may be easier, in the first draft, to use the personal pronouns "I," "we," "my," and then substitute less personal forms in the editing.

Examples:

First draft:	"My research indicates that . . ."
Second draft:	"The research indicates that . . ."
First draft:	"Based on the data, I feel that . . ."
Second draft:	"The data suggest . . ."
First draft:	"We were faced with three problems . . ."
Second draft:	"The investigators faced three problems . . ."

In doing a book review, refer to yourself as "the reviewer." Beyond this, however, be cautious when using the terms "the author" and "the writer." Both are considered by many editors as too self-conscious—although preferable to "I." (Note that as we have used "the writer," we have used it in a third-person sense.) However, both are acceptable in theoretical arguments and conclusion drawing. Such a prejudice against "the author" and "the writer" may seem irrational when such terms as "the researcher" and "the investigator" are regarded as respectable, but we are not here dictating to editors, only to those who must deal with them.

We note, however, that some fields seem to be accepting honest personal pronouns in scientific reports.

Since there is some divergence on this question, we recommend that you follow the practice of the agency or journal you are writing for.

The technical report is usually written in the past tense. As the editors of the American Psychological Association's manual of style point out: "The literature cited has already been written, the study's procedure has been carried out, and the results have been obtained."[3] Therefore, the body of the paper is centered on those things which have already taken place, and so the past tense is proper. However, within the report, some statements or sections should be written in the present tense. "A useful rule is that the present tense, in a research report, indicates statements which have a continuing or general applicability."[4] This would include definitions, hypotheses, and current theory.

Examples:

Definition: "Communication is [present tense, ongoing] the process of transmitting meaning between individuals."[5]

Hypothesis: "Consumption patterns operate [present tense, ongoing] as prestige symbols to define class membership, which is a more significant determinant of economic behavior than mere income."[6] (Hypotheses are also written in the future tense: "Consumption patterns *will* operate as . . .")

Theory: "Pure functionalism envisages them as forming mutually sustaining wholes [present tense, ongoing], but more recent authors have stressed [past tense, particular studies] the autonomy of institutions, and even that elements of the same institutions may be oriented to different ends."[7]

> Conclusion: "Oral literature collections by anthropologists and linguists usually offer [present tense, general finding] admirably recorded [past tense, particular studies] single versions of only an extremely small percentage of a repertoire of myths and tales."[8]

The social science writer usually follows conservative practice to make his writing appealing. Use of sensational incidents, mod diction, unorthodox sentence construction, emotional arguments, and such stylistic devices as excessive use of italics or boldface type for emphasis is regarded as cheap and undignified. Admittedly, such techniques are widely used in writings about the social sciences in popular magazines, but the social scientist who is writing in professional journals should maintain decorum and dignity in his writing if he wants to be accepted by his colleagues. This is not to say that he cannot subtly introduce new and useful techniques into his writing—this chapter will mention several—but he must recognize that change in the academic world is glacial.

In general, instead of using the rather flashy techniques just mentioned, the social science writer must depend instead upon contrast, emphasis, and sentence variation to make his writing appealing.

Contrast. The device of contrast is effectively used by social historian Richard Hofstadter to put across his point in the following paragraph.

> Where Burke is religious, and relies upon an intuitive approach to politics and upon instinctive wisdom, Sumner is secularist and proudly rationalist. Where Burke relies upon community, Sumner expects that individual self-assertion will be the only satisfactory expression of the wisdom of nature,

and asks of the community only that it give full play to this self-assertion. Where Burke reveres custom and exalts continuity with the past, Sumner is favorably impressed by the break made with the past when contract supplanted status: he shows in this phase of his work a disdain for the past that is distinctly the mark of a culture whose greatest gift is a genius for technology. To him it is only "sentimentalists" who want to save and restore the survivals of the old order. Burke's conservatism seems relatively timeless and placeless, while Sumner's seems to belong pre-eminently to the post-Darwinian era and to America.[9]

Emphasis. Among social scientists who study attitude change, primacy and recency are two common concepts. Just as important points should be placed at the beginning or at the end, whether it be a paragraph, section, or full report, the position of first and last should also be given attention in sentence construction.

Weak: The "West" began to evolve as a self-conscious section.

Strong: As a self-conscious section, the "West" began to evolve.

Emphatic: The "West," as a self-conscious section, began to evolve.

The third sentence was written by Frederick Jackson Turner. In his work, *The Significance of the Frontier in American History,* he effectively used this sentence—after a lengthy paragraph—as a short, emphatic, summary statement.[10]

Sentence Variation. Variety in length and structure of sentences is another device in making the whole paper both pleasing and effective. A sentence may be grammatically correct, but if it is followed by a long series of similar sentences, the monotonous repetition will surely tire the reader. Here is an example of sentence variation in a small paragraph.

> Let us go back a moment to the turn of the century. If we pick up the Protestant Ethic as it was then expressed, we will find it apparently in full flower. We will also find, however, an ethic that already had been strained by reality. The country had changed. The ethic had not.[11]

—Style as Format

The format for a technical report includes such things as placement and capitalization of headings, pagination, margins, indentations, abbreviations, spacing, and references.

Headings. In writing the report, the arrangement of the material on the page should reflect the organization of the material within the report. It is, in fact, a representation of the outline from which the writer has been working. Consequently, headings of equal importance should be given similar arrangement. Also the importance of the heading dictates the attention that should be called to it. Accordingly, major headings are generally centered on the page; minor headings are placed to the left.

The following is a standard form of headings for reports.

COLLECTIVE BEHAVIOR	(a)	Level 1 centered, all caps
Collective Behavior	(b)	Level 2 centered, upper and lower case, underlined
Collective behavior	(c)	Level 3 flush to left margin, underlined, initial cap only

Collective behavior.   (d) Level 4
indented five spaces, underlined, initial cap only

Collective behavior.   (e) Level 5
indented five spaces, initial cap only

For *a* (level 1), *b* (level 2), and *c* (level 3) headings, no period is used after the heading; the text follows on the next double-spaced line. For *d* and *e* headings, a period is used, and the text follows on the same line after a double space.

Some people concerned with technical reports insist that headings utilized in a paper must be in descending order; others insist that any suitable descending order is acceptable. Of course, if the paper includes five levels of headings, there is no option left to the writer. Few reports, however, have five levels. The following variations have widespread acceptance among writers and editors.

1. For reports with two levels of headings, use the *a* heading and the *c* heading, omitting the *b* headings, or the *b* and *c* headings.
2. For reports with three levels of headings, use either the *a*, *b*, and *c* headings, the *a*, *c*, and *d* headings, or the *b*, *c*, and *d* headings.
3. For reports with four levels of headings, use *a* or *b*, *c*, *d*, and *e* headings, or *a*, *b*, *c*, and *d* headings.

Often the sentences that form the sentence outline can be used directly as topic sentences when writing the report. This is particularly true in term papers and other less lengthy reports. For longer reports, on the other hand, the sentence outline may suggest too large an area to be developed in depth in one paragraph or two. In such cases, the sentence outline serves the same purpose as that of the key-word outline—to recall to your mind in proper sequence that which you want to develop in detail.

Here is the way the sentence outline from page 85 could be changed to headings in a short report. The descending order used is an *a, b,* and *c* sequence.

In Outline Form

I. Anthropologists tell us much about initiatory rites and ceremonies which accompany passage from one social state to another. These social states include passing from childhood to adulthood, from virginity to marriage, and from life to death.

   A. The purpose of all such rites is twofold. One purpose is to prepare the individual for the new way he should act. The second purpose is to instruct the members of the individual's social community in the new way they must act toward him.

   B. There are three phases in all rites of passage: (1) separation rites, (2) incorporation rites, and (3) transition rites.

      1. Rites of separation are prominent in funeral ceremonies.

      2. Rites of incorporation are prominent in marriage ceremonies.

      3. Rites of transition are prominent in ceremonies at birth.

   C. Often all three rites of passage are used in the same ceremony.

      1. A newborn baby may be separated from its mother or family.

    2. A newborn baby may be incorporated into the family or clan or tribe.

    3. A new born baby may be transmitted through baptism, or some other such ceremony, from a biological organism to a social being.

II. Differences in rites of passage ceremonies can be noted by contrasting a primitive society with a modern society.

In the Form of Headings When the Paper is Written

INITIATORY RITES AND CEREMONIES

Anthropologists tell us much about initiatory rites and ceremonies which accompany passage from one social state to another. These social states include passing from childhood to adulthood, from virginity to marriage, and from life to death.

(Other introductory material would go here.)

Purpose of Rites of Passage

The purpose of all such rites is twofold. One purpose is to prepare the individual for the new way he should act. The second purpose is to instruct the members of the individual's social community in the new way they must act toward him. (And other material.)

Phases in Rites of Passage

There are three phases in all rites of passage: (1) separation rites, (2) incorporation rites, and (3) transition rites. (And other material.)

Rites of separation

Rites of separation are prominent in funeral ceremonies. (And other material.)

Rites of incorporation

Rites of incorporation are prominent in marriage ceremonies. (And other material.)

Rites of transition

Rites of transition are prominent in ceremonies at birth. (And other material.)

Rites of Passage in Same Ceremony

Often all three rites of passage are used in the same ceremony. (And other material.)

Pagination. Place the page number—in arabic numerals—in the upper righthand corner four spaces from the top of the page and one inch from the right margin (or some similar consistent manner). Normally, pagination begins with the first page of copy; however, this will vary depending on the kind and nature of the report. While pages carrying major titles are counted, they are not given a typewritten number. Thus the second page in any chapter, major section, or one-unit report is the first page that has a typewritten number.

Pages are numbered consecutively to the end of the report, including pages of figures, tables, and the appendix, if one is included. Page numbers should appear alone; parentheses, hyphens, periods, or other decorative devices are not added.

If pages precede the introduction, they should carry page numbers in small roman numerals. The title page here, as well, should not be numbered, though it is counted as page i.

Margins and indentations. Most scientific writing requires minimum margins of one and a half inches on the left and one inch on the top, bottom, and right. Wider margins may be desired, and can be used. Margins must always be uniform from page to page, chapter to chapter.

Major sections of the report should begin on a new

page. It is suggested that they begin on not less than the fifth double space from the top of the page.

Five spaces is the customary paragraph indention.

Ten spaces is the customary paragraph indention on quoted passages of three or more lines. Subsequent lines are indented five spaces.

Abbreviations. In general, do not abbreviate words in the body of the report. To do so is to invite confusion and misunderstanding. Spell out the names of states, countries, months, days of the week, units of measurement; spell out all the parts of a proper name; spell out the words volume, chapter, and page; spell out titles when they are standing alone. Avoid the use of the ampersand (&) and other abbreviations such as *Bros.* or *Inc.* and *Ltd.,* except in copying official titles. When uncertain about forms of address in writing, consult a work such as *Webster's New Collegiate Dictionary* or another standard dictionary.

Abbreviations, on the other hand, are acceptable in tables, footnotes, charts, and the like. Within the body of the report, some abbreviations also are desirable. In this category are: Mr., Mrs., Dr., Messrs., Mmes. (as a title), St. (meaning saint, not street), and (after proper names) Jr., Sr., Esq., and degrees such as Ph.D., M.D., Ed.D.

If an organization, unit, category, or the like, such as the United States Department of Agriculture, Idaho State University, American Psychological Association, is referred to often in the report, it is acceptable to abbreviate to initials once the name has been given in full. If the organization or unit is not a common one, it is best to use such phrasing as ". . . hereafter referred to as USDA."

Wrong	*Right*
St.	street
Wm. James	William James
Hon. Snow	Hon. Sheldon Snow, or
	Hon. S. C. Snow
Rev. Pearce	Rev. Douglas Pearce, or
	Rev. D. N. Pearce
Dec. 16	December 16
chap.	chapter
The Dr. is in.	The doctor is in.
Utah Parks Co.	Utah Parks Company
We went down Mont. St.	We went down Montana Street.

Finally, when in doubt, spell it out.

Spacing. Type the technical report double-spaced, including headings. Normally, only three kinds of material are single-spaced in the body of the report: footnotes, quotations longer than three lines, and the bibliography section.

Double spacing is optional in tables and appendix material such as letters, questionnaires, and other tools used in the study.

References. Every reference cited in the report must appear in the reference list at the end of the paper. Each entry must be a complete identification. Perhaps the most widely used reference style is that of the University of Chicago Press as set forth in *A Manual of Style.* There is, of course, nothing sacred about any manual of style, but you need to be aware of the necessity of conforming to whatever style manual or sheet is prescribed by your supervisor, professor, or the publication you are writing for. As John A. Walter puts it, "Dance with the fella what brung you."

The purpose of the footnote and the bibliographical entry is to permit the reader to consult, should he desire to do so, the sources cited in your report, either to check the accuracy of data or to obtain additional information. This, then, calls for a complete citation. The style of the references, however, even within a discipline, will differ from journal to journal, from agency to agency. In some cases there may be only slight differences—as between *Sociological Inquiry* and *Urban Affairs Quarterly*, for example—but, nevertheless, the differences must be adhered to by the writer. This documentation style becomes part of the overall style of your paper.

—The Three Languages of Reports

We often assume that there is only one "language" used in writing—a native (in our case, English) language. Yet every society uses more than one system of communication, and there is much evidence to suggest that as a society becomes more industrialized the number of systems increases.

The social scientist as a technical writer habitually uses at least three systems: the language of words, the language of graphics, and the language of formulae. Further, the latter two are often superior to the first when the technical writer wants to communicate the special kind of knowledge which led to the creation of these two systems.

Each of these systems of communication has its subgroups:

    I. The language of words
        A. Conventional paragraphs
        B. Nonparagraph word uses
            1. Outlines
            2. Lists
            3. Word-based tables
            4. Readings and captions

II. The language of graphics
   A. Pictorials
      1. Photographs
      2. Drawings and paintings
      3. Maps
   B. Symbolic graphics
      1. Graphs
      2. Schematics
      3. Flow sheets
      4. Numerically-based tables
      5. Diagrams
III. The language of formulae
   A. Mathematical formulae
   B. Chemical formulae
   C. Statistics

Of these languages and sublanguages only the conventional paragraph language is treated as a communication device in conventional writing courses. Outlines are treated to some degree, but only as a device for improving rhetoric. Yet each of these latter two systems of communication has special value in conveying specific kinds of information. Indeed, if the social sciences are to move closer to the utopian state of one word–one meaning, one meaning–one word, more extensive use of the language of graphics and formulae needs to be made. By their use the writer comes closer to meanings which are singular and specific.

The Problem of Linearity. Spoken language is linear. That is, the voice can speak only one word at a time and the ear can hear only one word at a time. These words fall in sequence, or linear order. The listener has only his memory to relate what has gone before and only his imagination to predict what will follow. Written language is a little less linear than spoken language since the reader can look ahead or refer back (though conventional text does not encourage him to do so).

While this linearity is appropriate for narrative—especially suspenseful narrative—it is often a serious handicap to communication in other kinds of discourse. Consequently, the principles of rhetoric (that is, the principles of handling conventional paragraph language) were developed largely to overcome the limitations imposed by linearity.

(By way of example, many punctuation marks are signals of nonlinearity. Commas around nonrestrictive clauses, dashes, brackets, and colons are generally used to indicate that the linearity is changed either by interruption or in the case of series of items, by branching. Note also that the parentheses used here are a device for overcoming the limitations of linearity. Of course, most rhetorical transition words and phrases such as "furthermore," "on the other hand," "in conclusion," etc., are also signals relating to linearity.)

Not only the principles of rhetoric, but the systems of *graphics* and *formulae* self-evolved to overcome the handicaps imposed by language linearity. As a result, the technical writer has many forms to which he can turn to achieve greater clarity and precision in his writing. What follows is a brief discussion of the application of the several languages to five general areas in technical writing.

Process Description in Technical Writing. Since process description is essentially chronological, it is probably the technical writing technique best adapted to linear language. Yet even here it may be preferable to use the language of graphics (pictorials) to describe the hardware in the process and graphs to describe quantitative multifactor changes (time vs. temperature, for instance). Further, the steps of the process may often be made much clearer in numbered steps placed in a column than in conventional paragraphs. If the process is a chemical or mathematical one, the equation is a useful way of summarizing it in shorthand.

Spatial Description. This kind of description, also called mechanism description, apparatus description, or hardware description, is difficult to handle with the language of words, since the reader must not only visualize its relationship to what has already been described but also to what is yet to be described. (Perhaps the social scientist should be grateful that he does not often have to deal with it in his routine writing.) Such concepts as size, shape, color, hardness, texture, etc., in such a description can usually be handled in words only by relating the object or scene to some other object or scene in the previous experience of the reader. Obviously, pictorials (photos, drawings, blueprints, for example) simplify the task.

Definition. Definition is essentially an equivalency process, and the adaptability of the various languages depends upon the thing to be defined. A specific physical object is most easily defined by a pictorial. A process may also be defined better by the language of graphics or formulae. However, when the definition required is of an intangible or of a group, pictorials fail and only words will do.

Classification. Classification involves showing the relationship of the unit to the group. A near-ideal device for doing this is the outline—and not just the outline as an aid to rhetoric, but the outline as a communication device of its own. No other device so clearly shows componency—unless it is the exploded drawing which is a kind of outline itself. The basis of classification as well as individual items in the classification usually requires definition. For these purposes, the same criteria hold as for definition in general.

Interpretation or Evaluation. Interpretation is a comprehensive kind of communication. It involves most of the other techniques, and the kinds of language used for definition, process, and spatial description will apply here as well as when they occur separately. However, the comparisons involved in evaluation very often require the use of graphs, charts, tables, and mathematical formulae, since these allow simultaneous viewing of comparative data and thus easy demonstration of ranking and weighing of factors. General conclusions, however, are usually best shown in conventional language.

—Summary

For each writing situation you need to choose the language that will most clearly convey that kind of information to your intended reader.

In the past, there have often been strong prejudices against using graphs, tables, and other illustrative devices. Fortunately, those prejudices have largely been dispelled in the social sciences, though they still remain in some fields. The social science writer who does not become expert at using graphics is simply refusing to use a primary tool of his field. (Special texts on graphics are available in libraries and bookstores.) Clearly, when the information being presented involves such nonlinear concepts as shape, space, componency, simultaneous comparison, or quantitative relationship, the data can often be presented much more effectively in graphics, formulae, or nonlinear words.

Style in technical writing consists of two elements, the fashion in which words and sentences are put together and the following of certain prescribed procedures in laying out and typing the manuscript.

When putting words and sentences together, emphasize simplicity and clarity. The conventions of the stylesheet exist to help the writer put the material down on the page so that it will be logical, look attractive, and present the data in the most efficient fashion.

Notes

—Chapter 1

[1] See Ralph Borsodi, *Definition of Definition* (Boston, Mass.: Porter Sargent, 1967), p. 14.

[2] Ibid., pp. 11-12.

[3] Barbara W. Tuchman, "The Historian's Opportunity," *Saturday Review* 50, No. 8 (February 25, 1967), p. 27.

[4] Samuel T. Williamson, "How to Write Like a Social Scientist," *Saturday Review* 30, No. 41 (October 4, 1947), p. 17.

[5] Hans L. Zetterberg, *On Theory and Verification in Sociology* (Totowa, N.J.: The Bedminster Press, 1954), p. 33.

—Chapter 2

[1] William R. Parker, *The MLA Style Sheet* (Modern Language Association, 1951), p. 26.

—Chapter 3

[1] Morris L. Bigge, *Learning Theories for Teachers* (New York: Harper and Row, 1964), p. 183.

[2] Richard Hofstadter, *Social Darwinism in American Thought* (Boston: The Beacon Press, 1964), p. 11.

[3] Henry Nash Smith, *Virgin Land* (New York: Vintage Books, 1950), p. 51.

[4] Charles R. Wright, *Mass Communication: A Soci-*

ological Perspective (New York: Random House, 1964), p. 55.

[5] Donald M. McKay, *Information, Mechanism and Meaning* (Cambridge, Mass.: MIT Press, 1969), p. 139.

[6] Edwin Emery, *The Press and America* (Englewood Cliffs, N.J.: Prentice-Hall, 1962), p. 414.

[7] J. B. Lansing, R. W. Marans, and R. B. Zehner, *Planned Residential Environments* (Ann Arbor, Mich.: Survey Research Center, Institute for Social Research, University of Michigan, 1970), p. 109.

—*Chapter 4*

[1] William A. Scott and Michael Wertheimer, *Introduction to Psychological Research* (New York: Wiley, 1962), p. 337.

[2] Fred N. Kerlinger, *Foundations of Behavioral Research* (New York: Holt, Rinehart, and Winston, 1967), p. 39.

[3] Ibid., p. 14.

[4] Ibid., p. 23.

[5] Richard B. Braithwaite, *Scientific Explanation* (Cambridge, England: Cambridge University Press, 1955), p. 14.

[6] Pauline V. Young, *Scientific Social Surveys and Research* (Englewood Cliffs, N.J.: Prentice-Hall, 1956), p. 99.

[7] Adapted from Scott and Wertheimer. See pages 34-36 for fuller discussion.

—*Chapter 5*

[1] Lewis Mayers, *The American Legal System* (New York: Harper and Row, 1964), p. ix.

[2] William A. Scott and Michael Wertheimer, *Intro-*

duction to Psychological Research (New York: Wiley, 1962), p. 28.

—*Chapter 6*

[1] Karl Pearson, *The Grammar of Science* (London: Adam and Charles Black, 1911), p. 6.

[2] Pauline V. Young, *Scientific Social Surveys and Research* (Englewood Cliffs, N.J.: Prentice-Hall, 1956), p. 99.

[3] Bruce H. Westley, "Journalism Quarterly Style Book and Author's Guide," *Journalism Quarterly* 42, No. 2 (Spring 1965), p. 4, reprint.

—*Chapter 7*

[1] Bruce H. Westley, "Journalism Quarterly Style Book and Author's Guide," *Journalism Quarterly* 42, No. 2 (Spring 1965), p. 4, reprint.

[2] Ibid., p. 5.

[3] American Psychological Association, *Publication Manual* (Washington, D.C.: American Psychological Association, 1967), p. 17.

[4] Ibid.

[5] Charles R. Wright, *Mass Communications* (New York: Random House, 1964), p. 11.

[6] Pierre Martineau, "Social Classes and Spending Behavior," in James U. McNeal, ed., *Dimensions of Consumer Behavior* (New York: Appleton-Century Crofts, 1965), p. 173.

[7] G. Duncan Mitchell, ed., *A Dictionary of Sociology* (Chicago: Aldine, 1968), p. 101.

[8] Melville Jacobs, *Pattern in Cultural Anthropology* (Homewood, Ill.: Dorsey, 1964), p. 321.

[9] Richard Hofstadter, *Social Darwinism in American Thought* (Boston: Beacon Press, 1955), p. 8.

[10] Frederick Jackson Turner, *The Significance of the Frontier in American History* (New York: Frederick Ungar Publishing Co., 1963), p. 31.

[11] William H. Whyte, *The Organization Man* (Garden City, N.Y.: Doubleday Anchor Books, 1957), p. 16.

Index